A TRADITION OF INTEGRITY

A TRADITION OF INTEGRITY

THE STORY OF QANTAS ENGINEERING AND MAINTENANCE

BRUCE LEONARD

UNSW
PRESS

Published in Australia by
UNIVERSITY OF NEW SOUTH WALES PRESS
Sydney NSW Australia 2052
Telephone (02) 398 8900
Fax (02) 398 3408

First published in 1994

National Library of Australia
Cataloguing-in-Publication entry:

Leonard, Bruce, 1942– .
A tradition of integrity: the story of Qantas engineering and maintenance

Includes index.
ISBN 0 86840 110 2.

1 Qantas Airways — History. 2. Airplanes — Australia —Maintenance and repair — History. I. Title.

626.13460994

Typeset in 11/13 Weidemann Book, Quark Xpress
Printed by Southwood Press, Marrickville NSW

Available in North America through:
ISBS Inc
Portland Oregon 97213-3644
Tel: (503) 287 3093
Fax: (503) 280 8832

CONTENTS

FOREWORD

'Son! Do you know what you're doing?'

This simple, but profound, oft-repeated question came from Arthur Baird, Qantas' first engineer, who laid the foundations of engineering ingenuity and integrity in the fledgling airline. When I was a young engineer beginning a forty-year career with Qantas, I was deeply impressed by Arthur Baird's philosophy: 'Near enough is not good enough.' His influence persists even to this day and is largely responsible for Qantas' pre-eminent standing in world aviation.

In preparation for writing this book, Bruce Leonard interviewed more than fifty present and former engineering employees or their families, dating back to the mid-1920s. With his extensive research, involving volumes of written material (reports, memorandums, books, magazine articles), the author reveals a fascinating story of the engineers who contributed to the airline's growth and development. He recounts the struggles, trials and tribulations of the people who helped build the reputation that Qantas enjoys today.

As I look over the progress since the 1920s, I marvel that the company has survived so well in spite of many obstacles. The future presents even greater challenges, with Qantas' expansion into the domestic arena and the movement towards privatisation. Exciting technological advances are changing the whole nature of maintenance support of modern jet aircraft, but I have no doubt that the Qantas engineers will stay on top of it and Arthur Baird's philosophy of 'Near enough is not good enough' will continue to motivate their endeavours.

I thought I knew a lot about the history of Qantas Engineering and Maintenance until I read this book. It is interesting and easy to read, and it captures the spirit of the environment I enjoyed so much.

Ron Yates
(former Chief Executive,
Qantas Airways Limited)
September 1992

PREFACE

When the various histories of Qantas Airways Limited have been written in the past, there has been a perception that the contribution made by the Engineering Branch and its employees to the overall success of the airline as a whole has, in the main, been overlooked. This book is an attempt to pay tribute, not only to the pioneers of Qantas Engineering, whose work ethic created what I have termed 'A Tradition of Integrity', but also to the many hundreds of engineers who followed in that tradition during the company's first half-century.

It was never intended that I should attempt to write a complete 'history' of the Qantas Engineering Branch. My approach has been to relate the historic facts as known and available to me, with anecdotal evidence from various sources incorporated into the narrative to give readers a greater understanding of the 'human' face of the operation.

I should like to acknowledge and express my appreciation to many individuals, whose combined efforts helped me to locate and select material for inclusion in this book:

Ern Aldis, Jan Aldous, Sammy Ames, Sandra Avery, Phil Ballard, Bill Beard, Linda Beard, Bob Carr, Roy Carr, Mick Cook, Roger Critchley, Jack Drinkall, George Dusting, Eric Favelle, George Foord, Bruce Gates, Arch Graham, Bill Goodall, Ken Gould, John Gundry, Doug Hamilton, Mike Hards, Jeff Harris, Ruth Kydd, Jim Lamb, Pat Lord, Murray Mansill, Cec Morgan, Keith Ower, Norm Roberts, Frank Rowley, Doug Scott, Doug Stepto, Bob Struthers, Noel Tennant, Rona Thomson, Betty Upex and Bob Walker.

For their constant enthusiasm, guidance, encouragement and support in bringing this project to fruition, my sincere thanks to Bob Sprague (Qantas), George Roberts, John White, Ron Yates, John Burke (Qantas) and Mike Piquemal (Qantas).

To Christine Chang (Qantas), thank you for typing all my original drafts so graciously—and so often.

To my wife Jill (my live-in 'Thesaurus'), thank you for your continuing support and for always being there to provide a suitable word or appropriate phrase—and a welcome cup of tea.

And finally to Michael Ryan (former Qantas Director of Engineering and Maintenance), who had the foresight to recognise the necessity of initiating the project before it was too late, I say thank you for having the confidence in me to do the job.

I would also like to acknowledge the following reference sources used in the preparation of this book:

E. Bennett Bremner (1944), *Front Line Airline*
E.A. Crome (1955), *Qantas Aeriana*
Sir Hudson Fysh (1965), *Qantas Rising*, Angus & Robertson, Sydney
——(1968), *Qantas at War*, Angus & Robertson, Sydney
——(1970), *Wings to the World*, Angus & Robertson, Sydney
Barry Pattison and Geoff Goodall (1979), *Qantas Empire Airways (Western Operations Division) Indian Ocean Service 1943–1946*
John Gunn(1985), *The Defeat of Distance*
——(1987), *Challenging Horizons*
——(1988), *High Corridors*

Aircraft magazine (January 1930 and October 1961)
Empire Airways magazines (1934 to 1949)
QEA Staff News (1949 to 1953)
Qantas News (1954 to 1971)
Pix magazine (7 July 1956)

Bruce Leonard

Chapter 1

THE FORGOTTEN PIONEER

A number of books have recounted the beginnings in western Queensland of the small bush airline that grew to become Australia's international flag-carrier. Depending on which account of the Qantas story you read, credit for the successful operation of the fledgling airline is given mainly to Hudson Fysh, Paul McGinness or Fergus McMaster. While there can be no doubt that, each in his own way, these men played undeniably significant roles in fulfilling the company's original objectives, it should be acknowledged that there was a fourth man, a forgotten pioneer, the extent of whose personal contribution has until now gone unrecognised. He was one of the three original staff members of Qantas, a distinguished engineer whose singular expertise kept the company's old war-surplus aircraft operating during those early years. His name was Wilfred Arthur Baird.

Arthur Baird was born on 6 November 1889 at Benalla, a small town in northern Victoria. He attended Benalla West state school until he was sixteen years of age, when he moved with his family to the Melbourne suburb of Carlton. In Melbourne he completed an engineering apprenticeship and then began what was to be a lifelong association with aviation. After joining the Central Flying School at Laverton, he not only learned to fly but also saw nine months' service as an aviation flight instructor.

World War I was well into its second year when he enlisted with the First Australian Imperial Force in Melbourne, sailing for overseas service with No. 1 Squadron Australian Flying Corps in March 1916. He served in the Middle East, initially with the 17th Squadron Royal Flying Corps and then with the 67th and 111th Squadrons Australian Flying Corps. In 1918 he was promoted from Flight Sergeant to Chief Mechanic Flight Sergeant and awarded the Meritorious Service Medal for 'distinguished services in the field in connection with military operations culminating in the capture of Jerusalem'. He was also awarded the British War Medal and the Victory Medal.

While in Palestine, he won the respect and admiration of everyone who came in contact with him. His ingenuity with difficult aircraft engines often meant that a great deal of responsibility was thrust upon him. It is recorded that at Ramleh (in Palestine) in 1917, '... when No. 111 and 114 Squadrons were in dire trouble with their engines, he carried them through a very difficult time, and by his work on the fractious "Vipers", got their SE5s in the air again'.

It was during this period that Arthur Baird first met Lieutenant Paul J. McGinness, a talented and adventurous pilot, who was also stationed in Palestine with No. 1 Squadron. His observer was Lieutenant Wilmot Hudson Fysh, who had transferred to No. 1 Squadron from the 3rd Australian Light Horse Regiment. They enjoyed a very successful partnership and, with perfect understanding on tactics, close teamwork developed, resulting in both Fysh and McGinness being awarded the Distinguished Flying Cross.

The end of the war saw Hudson Fysh finally gain his 'wings' in Cairo, and in March 1919 he sailed for Australia on the *Port Sydney*, with Paul McGuinness and Arthur Baird also on board. After ten days in quarantine, Fysh returned to his home in Tasmania, but Paul McGinness decided to pursue his flying career. When Prime Minister William Morris (Billy) Hughes, in an attempt to encourage the growth of the infant aviation industry in Australia, announced a £10,000 ($20,000) prize for the first Australian to fly from Great Britain to Australia within twenty-eight days in an aircraft built in the British Empire, he decided to enter the race and wired Hudson Fysh to join him in the venture.

One of the battle planes that Fysh and McGinness had flown in Palestine was a Bristol Fighter, which had been presented to No. 1 Squadron by the wealthy grazier Sir Samuel McCaughey. McGinness therefore travelled to Yanco, in New South Wales, to see Sir Samuel. He presented their credentials of service, and the philanthropist agreed to provide the necessary backing for their entry in the air race.

With finance assured, Fysh and McGinness approached Arthur Baird in Melbourne, urging him to 'buy his discharge' and be part of the enterprise. At that time, Australian Military Regulation 327 (1916) stated: 'no soldier in the Permanent Forces can claim his discharge until the expiration of his term of service but ... he may, subject to the approval of the District Commandant and his Commanding Officer, purchase his discharge at the rate set forth below.' The rates varied from £20 for less than a year's service to £4 for more than four years' service. The cause of discharge under such circumstances was to be 'At his own request on payment of [rate] under Regulation 327.' Nevertheless, Arthur Baird's records show that he was discharged in Victoria on 1 June 1919 as 'Medically unfit' and not 'At his own request'.

Free now to participate in the air race, Baird made arrangements to travel to England with Fysh and McGinness. Baird was to be engineer, Fysh navigator and second pilot, and McGinness chief pilot. But shortly before their departure, Sir Samuel McCaughey suddenly died, and learning to their dismay that they had no claim against his estate, they were forced to abandon their plans.

By way of consolation, Fysh and McGinness were commissioned by the Department of Defence to survey a possible route for race contestants north from Longreach in western Queensland to the Gulf of Carpentaria and then on through Katherine in the Northern Territory, to Darwin. Arthur Baird was not included in this undertaking and returned to Melbourne, where he set up a small garage business in Lygon Street, Carlton, close to the family home. Fysh and McGinness set out on their adventure on 19 August 1919 in a T-model Ford fitted with a utility back and canvas hood, which had been specially prepared for the trip by Longreach Motor Works. They engaged George Gorham to accompany them as driver and handyman, for they knew it would be tough going for the greater part of the trip.

During the survey it was apparent to Fysh and McGinness that the vast, unconnected distances of western Queensland were perfect for an aeroplane operation. By August 1920, with extensive financial backing and support from grazier Fergus McMaster, they had raised sufficient capital to start an outback aerial taxi service: Western Queensland Auto Aero Service Limited. The name was subsequently changed to Australian Transcontinental Aerial Services Company Limited, but in November

1920 this name was also abandoned in favour of Queensland and Northern Territory Aerial Services Limited, the initials of which (coined with 'A.N.Z.A.C.' as its inspiring factor) provided the name Q.A.N.T.A.S. The company's headquarters were at Winton, the official birthplace of Qantas.

In a courageous move that was to change his life completely, Arthur Baird sold his garage business and joined Fysh and McGinness as the third full-time staff member of the new company, becoming its first engineer. Company records show that at the commencement of business, Paul McGinness and Hudson Fysh were both listed as 'pilots' and Arthur Baird as 'Aerodrome Manager'. As an expression of his faith in the future of Qantas, Arthur Baird bought an initial package of 50 £1 shares in the company (his shareholding would increase to 625 £1 shares by 1930).

Fysh and McGinness travelled to Sydney, where they met Arthur Baird on his way north from Melbourne. They contacted the head of the Australian Aircraft and Engineering Company, Nigel Love, at Mascot and ordered two Avro Dyak 504K converted wartime pilot trainers at a cost of £1500 ($3000) each. The original rotary engines on the 504K biplanes were unsuitable and were being replaced with water-cooled 100 horsepower vertical six-cylinder Sunbeam Dyak engines (produced to power the dirigibles, for which they had given good service). But delays in the delivery of one of the engines resulted in Qantas arranging to take only one Avro 504K. In place of the second aircraft they ordered an Avro 547 triplane, which would carry four passengers in an enclosed cabin, at a cost of £2798 ($5596).

Qantas took delivery of its first aircraft, the Avro 504K (G-AUBG), on 30 January 1921. Of wood and fabric construction, the 504K had a cruising speed of 65 miles per hour with the pilot and two passengers. It was 28 feet 11 inches long and had a wingspan of 36 feet.

While waiting for the Avro to become available, Fysh and McGinness were approached by Charles Knight, a leading stock and station agent in Longreach, to fly the BE2E aircraft he had purchased from the Perdriau Rubber Company, to Longreach. They readily agreed, and on 31 January 1921 the two aircraft took off from Mascot with Paul McGinness piloting the Avro and carrying Charles Knight as a paying passenger and Hudson Fysh at the controls of the BE2E with Arthur Baird in the passenger seat. It is interesting to note that, at this time, Paul McGinness was a very experienced pilot with some 500 hours to his credit. Arthur Baird, too, had accumulated many hours of flying time in various aircraft types since his pre-war days at Laverton. But Hudson Fysh had only 34 hours and 50 minutes flying experience, which included a mere 30 minutes in a BE2E. Despite this, for reasons that are not recorded, he elected to pilot the BE2E to Longreach himself and carry Arthur Baird as his passenger.

The journey north was not without incident. Flying along the New South Wales coastline they ran into a mass of dark cloud south of Newcastle. Neither aircraft was equipped for 'blind flying'. The only instruments carried were a magnetic compass, an altimeter, an inclinometer, a tachometer for engine revolutions and an airspeed indicator. Stalling speed on the BE2E was only 25 miles below top speed. McGinness was able to direct the Avro through gaps in the thickening cloud, but as the BE2E began to be tossed about by turbulence, Fysh became disoriented. The engine stalled, and with the nose swinging out of control the aircraft started into a spin. Hudson Fysh described what happened next:

> 'With the first turn of a fatal spin commencing, out [of the clouds] we came with a swish into a small clear patch in a valley, but with the clouds on the ground all round. To right the stable old BE2E was only a matter of moments. On my right were the buildings and pithead structure of a coal mine, with a sloping, bushy hillside alongside. I promptly put the BE2E down on the hillside, careered uphill through the bushes, and came to a halt without damage, beside a miner's cottage.'

Incredibly, neither Fysh nor Baird were hurt, and the BE2E sustained no damage. The coal mine was at Red Head, and as weather conditions improved they took off again, to be reunited in Singleton with an extremely anxious Paul McGinness. From Singleton to Longreach the flight was uneventful. They were met at Barcaldine by Fergus McMaster, who climbed aboard the Avro for the 65 mile flight and triumphant arrival at Longreach.

It did not take long for the owner of the BE2E, Charles Knight, to realise that he had made a mistake in buying the aircraft. On a particularly rough and difficult flight with Hudson Fysh between Winton and Longreach, during which Knight became very sick, he vowed he would never fly again. As a result of this decision, the Qantas fleet doubled with the purchase of the BE2E from Charles Knight for £450 ($900).

The BE2E biplane (G-AUBF) was powered by a 90 hp Royal Aircraft Factory engine and cruised at 65 miles per hour fully loaded. It went into service with Qantas in February 1921, when Hudson Fysh began joy flights and charter work from various western Queensland towns. Herbert Avery, proprietor of the Western Motor Works at Longreach, was often taken along as 'honorary' mechanic on these flights between towns.

Local support for 'the aeroplane people' (by which name the Qantas fraternity became known) emerged from a number of sources. The Winton Shire Council offered to pay half the cost of preparing a landing strip, to a maximum of £10 ($20), and the Alba Wool Scouring Company made one of its Winton wool stores available to Qantas to house its aircraft free of charge. When the company shifted its head office from Winton to the now-familiar Duck Street office in Longreach (about 160 kilometres southeast of Winton), the Longreach Agriculture Society made space available near the showground, which became the site of the first Qantas maintenance hangar. Built at a cost of £117 ($234) it was little more than an open-fronted nose hangar, but it remained the base of the company's maintenance operation until August 1922, when the new hangar at Longreach aerodrome was ready for use.

Meanwhile, Paul McGinness and Arthur Baird had gone to Sydney to take delivery of the third Qantas aircraft, the Avro 547 triplane. Unfortunately, McGinness severely damaged the aircraft during trials, shearing off the undercarriage while coming in for a practice landing under full load. The propellor, radiator and one wing tip were also impaired during the resultant crash landing — as a consequence of which, the aircraft's certificate of airworthiness was permanently withdrawn. Almost half the capital of the company had been invested in an aircraft that never went into service.

In late 1921 the Australian government invited tenders to operate a subsidised air mail and passenger service between Charleville and Cloncurry. Qantas was awarded the contract, against heavy opposition. Although it would be twelve months before the first scheduled service, it was obvious that the company's existing fleet and facilities needed to be upgraded. Two eight-seater Vickers Vulcan aircraft were ordered from England to operate the service, and a contract was signed with Stewarts and Lloyds in Brisbane for the supply and erection of the hangar at Longreach aerodrome for £1637 ($3274).

Since the start of operations, Arthur Baird had been 'toiling with Trojan vigour' as the company's sole engineer, keeping the fragile 504K and BE2E flying. Now, with additional aircraft on order and regular scheduled services due to begin, it was realised that he would no longer be able to do all the mechanical work himself. During 1922 his engineering staff was increased to four with the employment of Frank McNally (mechanic), George Boehm (woodworker), Jack Hazlett (mechanic) and George Dousha (woodworker). A Talbot utility was bought for £275 ($550) to carry spares and stores between aerodromes and emergency grounds, and Stewarts and Lloyds also started building hangars at Charleville and Cloncurry. The Shell Oil Company began its long association with Qantas by establishing petrol and oil depots at all the major centres along the 570 mile route, each providing about 1 ton of fuel.

By April 1922 industrial disputes in England were signalling substantial delays in delivery of the Vulcans, and Qantas was forced to seek substitute aircraft so that it could start the air mail and passenger service. Fysh therefore bought a De Havilland DH4 biplane in Melbourne from Ray Parer (who, with John McIntosh, had been the first to fly from England to Australia in a single-engined aeroplane in 1920), paying £1500 ($3000). The DH4 (G-AUBZ) was sent by rail to Longreach, but it arrived damaged on 12 August and had to be completely dismantled by Arthur Baird in the newly completed Qantas hangar. The aircraft's logbook details the extent of the modification work carried out by Baird and his small team:

> 'The cabin was removed and ring bracings placed in the fuselage; new compression struts were attached and a new top section constructed and fitted. Completely new ply and fabric covering was fitted to the fuselage and the undercarriage was completely dismantled. A new tail skid was fitted and shock absorbers renewed throughout the machine. The engine bearers and fuselage were tuned up and Rolls-Royce engine No. 158 was installed.'

It was described as a 'clumsy machine, with its four-bladed propellor and its steel ladder on the outside of the passenger cabin,' but its 360 hp Rolls-Royce Eagle VIII engine enabled it to cruise at 90 miles per hour when fully loaded with a pilot and two passengers. The DH4 was to have a long and useful life with the company.

Qantas then tried to obtain two DH9s from the Royal Australian Air Force pending the arrival of the Vulcans, but were unsuccessful. Things looked grim until the Federal Government stepped in and offered to assist with the purchase of two Armstrong Whitworth FK8s for £700 ($1400) each. Despite his increasing workload, Arthur Baird was sent to Melbourne to inspect the Armstrongs and arrange for their shipment to Longreach. They arrived in early September and were assembled and test flown by Fysh and McGinness, with Arthur Baird as passenger. The FK8s (G-AUCF and G-AUDE) were each powered by a 160 hp six-cylinder Beardmore engine and could carry a pilot and three passengers. The Beardmore engine from the discarded Avro triplane was shipped from Sydney and used as a spare for the FK8s.

Although tests on the aircraft had shown them to be sound, the Armstrong Whitworths had not been designed to operate in tropical inland conditions under full load. Their rate of climb, only 500 feet per minute, meant that they had to be coaxed off the ground in hot weather and then kept in the air 'with an excessive throttle opening'. This caused the radiator to boil, resulting in the loss of a large amount of water and a number of forced landings. Baird overcame the problem by replacing the cool-climate radiators with larger ones more suited to harsh outback conditions. He also installed a header tank which, by condensing the steam, conserved the water in the system.

Charleville aerodrome, 22 November 1922, 5:30 am: The Qantas Armstrong Whitworth FK8 G-AUDE stood fuelled and ready. A small crowd cheered as Paul McGinness climbed into the cockpit. Arthur Baird swung the propellor, and the Beardmore engine roared to life. Carrying 108 letters, the FK8 took off on the first leg of the Qantas inaugural service as a scheduled airline operator, with Arthur Baird travelling as engineer. After a stop at Blackall, they arrived at Longreach ahead of schedule after only four and a half hours' flying time, to be welcomed by Fergus McMaster and a crowd of townspeople. Because of the intense heat of the midday sun in western Queensland, Longreach was made an overnight stop.

Soon after sunrise the next morning, the second FK8 (G-AUCF) was loaded in preparation for the second leg of the historic flight to Cloncurry. Hudson Fysh was to pilot this sector, with Arthur Baird continuing as flight engineer. The flight also carried Qantas' first paying passenger, eighty-four-year-old Alexander Kennedy, one of the original subscribers to the company. But the aircraft's Beardmore engine refused to provide sufficient revs, and after three failed attempts at take-off, they transferred to the other FK8. When the flight finally got under way 30 minutes late, Alexander Kennedy, in cap and goggles, with the wind tossing his beard, shouted 'Be damned to the doubters!' as the FK8 left the ground.

The Qantas story had begun.

A series of setbacks marked the beginning of 1923. Paul McGinness, unable to conform with the disciplines of a scheduled airline operation, resigned and left the company. Then, with the new service barely four months old, Hudson Fysh crashed one of the FK8s (G-AUCF) during a mercy flight at Jericho, completely destroying the aircraft and depleting the company's already-meagre fleet.

The fact that the four remaining aircraft were antiquated and yet managed to keep to the weekly schedule gave heart to the Board of Qantas, as it emphasised the excellent ground organisation and the skill and technical integrity of Arthur Baird. On 23 June 1923, Hudson Fysh, in a memorandum to Fergus McMaster (who had been elected chairman of the Qantas Board), was moved to report the following:

> 'During the eight months this service has run, our mechanical staff have been right up against it from the start.'

He went on about the conditions: the aircraft were second-hand to start with, and only one was suitable for summer flying. The men worked long hours, including Saturdays and Sundays. The report continued:

> 'Meanwhile the work of carrying on is becoming more of a problem and, when one considers the indifferent and understandardised plane which had to

> be kept airworthy and efficient, I consider that no other mechanical staff would have produced the same results.
>
> I claim that our staff, under Mr Baird, produces as good results with second-hand machines as we could have expected from a new plant.
>
> Colonel Brinsmead [the first Controller of Civil Aviation in Australia] told me recently that we had the best mechanical staff in Australia, and we were fortunate to have the services of Mr Baird.'

Meanwhile, the long-awaited Vickers Vulcan had finally arrived. It was labelled the 'flying pig' because the ungainly shape of its fuselage made it appear fat, which in turn caused it to lumber along the runway resembling a sow on the run. But the Vulcan was found under test to be unsatisfactory and not in compliance with the contract, so Qantas refused to accept the aircraft.

This called for desperate measures to be taken, if only as a stop-gap. A Bristol Fighter F2B was purchased, and two De Havilland DH9Cs were ordered from the British parent company for delivery later in the year. The Bristol Fighter F2B biplane (G-AUEB), bought from aviation pioneer Horrie Miller for £1000 ($2000), had a fabric-covered airframe and could carry one passenger and the pilot at a cruising speed of 90 miles per hour. It went into service immediately and gave superb performance even in the hottest conditions. Its 300 hp Hispano-Suiza V-8 engine was water-cooled, but it was lubricated with pure castor oil, the fumes from which would invariably envelope the pilot and cause the passenger to become sick.

In June 1923 Hudson Fysh flew the Avro on a charter job to the races at Windorah on Cooper Creek, with the Longreach bookmaker and his penciller on board. During the return trip to Longreach, heavy weather forced them to land at a property called Guildford Park on the Thompson River, where the ensuing gale-force winds and torrential rain damaged the aircraft.

Help was needed from Longreach to effect repairs, so pilot Fred Huxley flew Arthur Baird out to Guildford Park in the DH4. As Huxley was coming in to land, he flew through a telephone wire that was strung on bush poles along one side of the landing area. The wire broke and began to wind itself around the propeller hub. The DH4 careered on, and as the No. 8 gauge wire became taut, the aircraft came to an abrupt mid-air stop, crashing heavily to the ground. Miraculously, Huxley and Baird stepped out of the wreckage unhurt, but the DH4 had to be trucked back to Longreach for repairs.

The company's fleet was strained almost to breaking point as the year entered its second half, with the remaining Armstrong Whitworth and the Bristol Fighter left to operate the service, and the Avro and BE2E on standby for taxi and charter work. But the worst was not over. On 13 September the Armstrong experienced engine failure near Blackall, and in the forced landing that followed, the undercarriage, wing and propellor were damaged. The aircraft could not be flown and had to be railed back to Longreach. In desperation, the company bought a third Armstrong Whitworth FK8 (G-AUCS), which joined the Bristol Fighter on the service in October 1923.

The same month, the Government renewed the Qantas subsidy contract to fly the Charleville–Cloncurry air route for another twelve months. The two DH9Cs (G-AUED and G-AUEF) arrived from England in November 1923 and were assembled at Longreach. Both proved satisfactory when tested by Fysh but were modified by Arthur Baird before being placed into service to give the passengers the comfort of an enclosed cabin. The DH9C was a converted wartime biplane with slightly swept-back wings, a design feature that altered the centre of gravity and enabled it to cope with more weight astern. Three passengers and luggage could be accommodated, with enough spare weight allowance for some small parcels and a few pounds of mail. Its 240 hp Siddeley Puma engine enabled it to develop a cruising speed of 82 miles per hour.

The year 1923 had been a period of mixed fortunes for Qantas, and without the dedication and engineering skills of Arthur Baird and his staff it seems certain that the company would have been forced to cease operations. During the year a further increase in the engineering staff had been deemed essential and resulted in Phil Sullivan (woodworker), F. Wellington (mechanic), E. Ellis (woodworker) and Pat Andrews (woodworker) being employed.

With the introduction of the DH9Cs, the situation seemed brighter. Qantas was at last in a position to operate its air passenger and mail service on a substantial basis. But before the year was out, another crash would put the Bristol Fighter out of commission for almost twelve months.

It seems that the Bristol was a difficult aircraft to fly, and on 18 December 1923 a pilot named Vigers, who was new to the company, was given permission to fly the F2B around Longreach to familiarise himself with the controls. He was inexperienced with the aircraft type and quickly got into trouble, sending the Bristol into a spin and crashing it nose-first into the ground just behind the railway station. Damage to the aircraft was severe, but it too was salvaged, rebuilt as a Bristol 'Tourer' and eventually returned to service by Baird and his engineers. The Hispano-Suiza V-8 engine was found to be beyond repair and was replaced with a Siddeley Puma engine (which also powered the DH9Cs). Another improvement to the original design saw the rear cockpit fitted with a cover. Rebuilt, the Bristol gave good service as a reserve aircraft until it was finally sold in 1927.

Vigers, in contrast, did not have a long and illustrious career with Qantas. He was fired shortly after the Bristol incident, having been reported for putting one of the DH9Cs into a loop on the Cloncurry–Longreach sector of the run with two elderly female passengers on board.

CHAPTER 2

A WORLD FIRST

In February 1924 the Qantas Board gave approval for the purchase of a De Havilland DH50 aircraft, the first aeroplane type built specifically for civil work to be used by an Australian airline. Compared to the aircraft that Qantas had operated to this time, the DH50 was a luxury airliner with a genuine enclosed passenger cabin, which could carry up to four travellers (the pilot sat outside in an open cockpit, braving the elements).

The company's first and only imported DH50 (G-AUER, named *Hermes*) was introduced into service in October 1924. It was powered by a 240 hp Siddeley Puma engine and in good conditions could achieve a rate of climb of 600 feet per minute and a full-throttle speed of just over 80 miles per hour. Take-off performance was improved significantly by the fitting of a four-bladed propeller by the Qantas engineers at Longreach, and fully loaded the DH50's range was 375 miles. The arrival of *Hermes* marked the beginning of a new era for Qantas. It proved to be a great success and became an aircraft of great historical importance during the ensuing years. Shortly after its entry into service it came to the rescue of Prime Minister Bruce and his wife, flying them from Winton to Longreach when a storm cut all roads in the district and put an end to their motoring tour of western Queensland.

Incredible as it might seem, the damaged DH4 was also brought back into service, having been completely rebuilt and test flown satisfactorily by Arthur Baird's team at Longreach. A new cabin had been designed and installed (after the style of the DH9Cs), which greatly improved the appearance of the aircraft, and longer exhaust pipes had also been fitted to cut down on noise. Both improvements added to passenger comfort.

There can be no doubt that Arthur Baird was the backbone of Qantas in those early, uncertain years. He had built up a skilled staff of engineering specialists who dealt quickly and efficiently with the constant influx of damaged aircraft and failed engines. With such expertise at its disposal, the Board made the decision to embark on a venture that would make history in aviation circles: Qantas was to become the first airline in the world to build and fly its own DH50 aircraft. It was also an undertaking that had no parallel in Australia at the time. The agreement allowed for aircraft to be constructed in the Qantas hangar at Longreach, under licence, with the payment of an initial royalty to De Havilland of £100 ($200) per plane. This was later reduced to £50 ($100).

The Longreach hangar now had a concrete floor and a raised timber area at the rear, where the modest but adequate woodworking equipment was housed. It had also been linked to the town's electricity supply and was ready to face the new challenges of the months ahead. The engineers were all specialist tradesmen, the best of whom had served their time as fitters and turners and pattern-makers. Blueprints were spread out on long trestles and studied closely; jigs were set up, and saws began to buzz, amid the pungent smell of dope from the canvas-screened doping area on days when the temperature and humidity allowed its application.

The engines, tanks, radiators, instruments and all-metal parts were imported from England. Spruce and Oregon timbers for wing spars and longerons were Canadian, but the maple used for propellers and the three-ply needed for cabin construction came from Queensland. Arthur Baird was meticulous in his choice of spruce logs in Brisbane and in getting three-ply made to the exact thickness and specification from the supplier, Bretts. Irish linen for wing coverings came in large rolls from Ireland, while the dope (to make the linen taut) came from England.

It was important to Arthur Baird that his construction staff at the aerodrome work in a happy environment, and he introduced music into the hangar (presumably by means of a wind-up gramophone) to pep up the work and keep the pace going. Hudson Fysh records his amusement at the sight of Bill Stone and Neville Hagarty providing their own musical rendering, as many a gimp pin and screw was driven home to some popular tune whistled or sung by the two. Apparently, Bill Stone's star turn was a xylophone solo performed on the rib work of an uncovered DH50 wing with the aid of two tack hammers.

Not a lot was known about the individual people who contributed to the success of the Qantas DH50 project, until the opportunity came recently to speak to the only surviving member of Longreach DH50 construction crew, eighty-six-year-old Bob Struthers. Bob joined Qantas in 1927 after answering an advertisement in the Brisbane *Courier Mail* for 'two specialist woodworkers, preferably cabinet makers or shop fitters, for Qantas at Longreach'. He was twenty-two years old at the time and had served an apprenticeship at his uncle's furniture business in Scotland while attending evening classes at Glasgow Technical College. These were the days when the woodworker was supreme.

George Boehm (who had started Qantas off on its woodworking tradition in 1922) had left, and Bob Struthers was employed to take his place. Bob arrived at Longreach after a 36-hour train journey from Brisbane, having travelled with the other successful applicant, Bill Stone. It is interesting to note Bob Struthers' first impression of Longreach, having been used to the lush hills of Scotland: 'The only green grass I saw was around the War Memorial.' The next morning they were taken out to the aerodrome on the back of the Talbot utility and were met by Arthur Baird. Bob Struthers was given timber to construct his own workbench on the raised section at the rear of the hangar, and he set up next to George Dousha, an American of Swedish descent, who had been a woodworker with Qantas since 1922. Together they constructed the DH50's wings, while Pat Andrews manufactured the auxiliary parts (struts, etc.) for the wings. Bill Stone worked with Neville Hagarty, a local Longreach man, on the main body structure of the aircraft and also on the cabin interiors. Although each man was a specialist in his own field, everyone got on well together and it was a very happy working environment. According to Bob Struthers, the spruce timber was lovely to work with; it was light, easy to handle and strong, but most importantly it kept straight. The woodwork section had three machines — a planing machine, a spindle for gouging out (a routing machine) and a band saw — although the woodworkers were not allowed to operate them. Because they were working on a very low budget and the timbers were expensive, there could be no room for error, so Arthur Baird would operate the machines himself. He was apparently not only a top engineer but also a very competent machinist. Bob Struthers was sometimes permitted to drag the timber off the machine.

Working on the wings (from blueprints supplied by De Havilland) required very intricate and fine workmanship. Once the frames were complete, they were covered with Irish linen by a labourer in the hangar, Bill Tucker, who would then apply the dope to shrink and tighten the stretched fabric.

Arthur Baird oversaw the entire construction process. He would always check progress of the work, particularly the line of the wing to make sure it was straight, after which he would simply say 'very good' and walk on. It was not his style to rebuke or admonish any of his workers; he preferred to encourage them and keep their spirits up. In the workshop vernacular, Arthur Baird was known as 'Beeno'. Why this nickname was ascribed to him, no-one has been able to recall, but he is reported to have still been referred to by this name well into the 1930s. He had a pet dog called Kelly, which accompanied him wherever he went. Kelly was a black and white fox terrier and had been trained to sit on the bonnet of Baird's old Hudson car as he drove to and from the airport each day. The Hudson was never driven faster than 25 miles per hour and, with Kelly perched up on the bonnet, the two became a familiar sight around Longreach.

Although not often seen, Arthur Baird's sense of humour did emerge when the occasion called for it. On one particular day, Bill Stone was working on the assembly of a DH50 fuselage. This was done in jigs, which held the three-ply to be glued, screwed and bradded ('brads' being small nails with no heads). When fitting a sheet of three-ply it was essential that all the brads be tacked into position before the glue set, and Bill Stone, being ambidextrous, could tack up one row of brads, then switch the hammer to his other hand and tack along in the opposite direction. Observing this for the first time, Arthur Baird was most impressed and suggested to Bill that if he used two hammers simultaneously he could do twice as much work in the same time. Bill is reported to have answered, 'Sure, that will be fine, but I'll also want twice the pay.' No record can be found of Arthur Baird's reply.

Once all the component parts were completed, it took between two and three weeks to assemble a DH50. To ensure continued quality of workmanship during the entire construction process, Arthur Baird introduced a unique but effective incentive scheme: tradesmen who worked on an aircraft's construction were always the first to fly in it during test flights. Needless to say, the craftsmanship of the Qantas-produced DH50s at Longreach was seen to be as good, if not better, than the aircraft built by De Havilland in England.

In 1926 Major Geoffrey De Havilland came to Longreach to check on the progress of the DH50 building program. The first Qantas DH50'A' (G-AUFA) rolled out of the Longreach hangar on 18 August, and everyone associated with its construction dressed in their 'Sunday best' to pose in front of the aircraft for the official photographs. It was christened by Lady Stonehaven, wife of the governor-general, who proclaimed that it should be known as *Iris*, a sure-footed messenger of the gods and personification of the rainbow. After the ceremony Lord and Lady Stonehaven boarded the plane for a long flight to Newcastle Waters. Then Hudson Fysh loaded the Talbot utility with everyone who had worked on the new aircraft and drove them to the Imperial Hotel, where he bought several rounds of drinks to celebrate the success of the venture. This became a tradition enjoyed by the engineering staff each time a new Qantas-built aircraft entered service.

A further boost to the construction staff's morale came in the form of a letter received by Hudson Fysh:

'Melbourne
27 Aug, 1926

Dear Sir,
I desire to congratulate your Management on the very satisfactory report submitted by our Departmental Inspector who has just returned from Queensland.

This report is the most recent of a series which all express great commendation of the technical efficiency of your service management, and it is a matter for mutual congratulation that your painstaking attention to details is so admirably reflected in your service record.

Particularly do I wish to convey my appreciation of the success which has attended your initial venture in local aircraft construction, as I am informed that the new De Havilland 50, constructed almost entirely at Longreach by your Staff, has proved satisfactory on test and bears comparison in every way with the imported machines.

I am confident that the success which has attended this undertaking will prompt your Management to further efforts in this direction, in which you have my very best wishes for success.

Yours faithfully,
H.C. Brinsmead

Controller of Civil Aviation'

Lester Brain, a cadet from Point Cook, had been the first non-wartime pilot to join Qantas in 1924, and Dudley Wright (who was eventually to succeed Arthur Baird as Chief Engineer) was employed as a mechanic in June 1927. The following month Arthur Baird started his now famous Qantas apprentice training scheme at Longreach with the appointment of fourteen-year-old Jack Avery (son of the local garage proprietor Herbert

Avery) as its first apprentice, although he was listed on the company's records as 'Boy at Aerodrome'. This proved to be a fortunate appointment, for by the time he retired in 1972 after a distinguished engineering career of some forty-five years, Jack Avery was the company's longest serving employee. His contribution to the development of Qantas was recognised in January 1967 when he was awarded the MBE (Member of the British Empire, Civil Division) for his services to civil aviation.

With the success of *Iris*, construction of a further two DH50As was started. However, before these were completed, a cabin DH9 aircraft (G-AUFM) was built and named *Ion* by Arthur Baird. It emerged from the hangar in February 1927, but despite the fact that it flew nicely and had a very good cabin that could accommodate three passengers, it was not used extensively because of the rapid introduction of additional DH50A aircraft. The second Qantas-built DH50A (G-AUFU, named *Perseus*) began operation on 11 June 1927, and the third (G-AUGD, named *Pegasus*) flew its first scheduled service less than two months later on 2 August 1927.

Although the Puma-engined DH50A proved to be a splendid aircraft for Qantas operations, it was found to have no performance in reserve for the harsh conditions of western Queensland. On Arthur Baird's recommendation, a decision was taken that resulted in Qantas building three DH50'J's, which were powered by the larger-capacity 450 hp Bristol Jupiter Mark VI radial engines. The substitution of the Jupiter engine in place of the Puma almost doubled the horsepower (from 240 to 450 hp), as well as the cost of producing the aircraft. This cost disadvantage was not offset by any increase in payload or volumetric capacity — the DH50J still carried only four passengers and the pilot, with a small amount of room for luggage and urgent freight. However, it was balanced by a nearly doubled rate of climb (from 600 to 1000 feet per minute), an increase in cruising speed (from 80 to 105 miles per hour) and greater range with the extra fuel tanks that were fitted. The Jupiter engine proved far more reliable than the Puma, and great cost savings resulted from the increased periods between overhauls. (After nine months of operation the first Qantas-built DH50J had flown 26,000 miles in 348 hours. Although the engine was running perfectly, it was removed for its scheduled overhaul. At that stage, recorded replacement costs had been less than £1 ($2), providing a dramatic contrast with the Siddeley Puma engine, which rarely achieved 75 hours without needing serious attention.)

The first Qantas-built DH50J (G-AUHE, named *Atalanta*) was flown on 30 March 1928, followed by the second (G-AUHI, named *Hermes*) on 12 May 1928. Then in 1929 another DH50A (VH-UJS, unnamed) was built, and in August the last DH50J (VH-ULG, named *Hippomenes*) entered service.

Qantas had also sought permission to build DH60 Moths, but De Havilland refused, and since the DH61 Giant Moth was too large an aircraft for the company to build at its Longreach facility, the Qantas building program ceased with the completion of *Hippomenes* in 1929.

In his book *Qantas Rising*, Hudson Fysh attributes the success of the Qantas aircraft-manufacturing venture to the devotion and ingenuity of Arthur Baird and his team of engineers. It was an outstanding achievement, particularly when it is realised that the building projects had to be sandwiched between normal maintenance of aircraft and the continued incidence of damaged aircraft that had to be lorried or railed back to Longreach for repair.

Because Longreach was in such an isolated area of western Queensland, Qantas had to pay top wages to get qualified tradesmen to come and work there. Bob Struthers recalls that the average wage paid by Qantas to its specialist engineering staff was £6 ($12) per week, which in the depression conditions of the late 1920s was considered exceptional recompense for the discomfort of living and working in the outback's oppressive climatic conditions. Most of the men stayed at a boarding house in town and spent evenings enjoying a few drinks at Mum Springer's at the Imperial Hotel. The main sporting activity was tennis, and the only regular entertainment (apart from the odd travelling show) was the Saturday night picture show. It took place in an open-air theatre owned and operated by a Chinese named Fred Afou, whose daughter played the piano to accompany silent movies.

By the mid-1920s Qantas could foresee the coming era of private ownership of aircraft and a need for training pilots for private licences. As there were no flying schools in Queensland, Qantas set out to establish its own, using the eminently suitable DH60 Moth aircraft which had just become available from De Havilland. Four DH60s with Cirrus engines were purchased (one of which was assembled at Longreach), and Qantas was in the flying school business. The first school opened to a zealous response at Longreach in December 1926 with pilot Charles Matheson as the main instructor. But the initial gusto soon petered out in the face of drought conditions, coupled with a number of crashes and the difficulty of instruction in the gusty western Queensland summer air.

Arthur Baird went to Brisbane to select a site, build a hanger and start a company flying school there, and the Brisbane Flying School opened at Eagle Farm in March 1927 with the splendour of an aerial pageant. Lester Brain was sent to Brisbane to manage the operation and was its first instructor. He was later joined by Russell Tapp and C.W.A. (Charles)

Scott as the school began to grow and flourish. It was eventually handed over to the Queensland Aero Club in May 1929, after the training of fifteen pilots to 'A' licence standard, although Qantas continued to carry out the instruction of pilots and maintenance of aircraft under contract to the Club.

Meanwhile, the Qantas apprenticeship scheme began to gather momentum with the appointment of Bill Bennett as the company's second apprentice at Longreach on 10 December 1928. Then, on 27 August 1929, fourteen-year-old Eric Kydd began his apprenticeship at Eagle Farm. Although he was officially the company's third apprentice, he was in fact the first 'indentured' apprentice, but like his predecessors he too was listed on the company's records as 'Boy at Aerodrome'. This is not altogether surprising when you consider that his sole duty initially was to wash the company's aircraft. Both Bill Bennett and Eric Kydd would spend their entire working lives with Qantas and assume important positions in engineering as the company developed.

In May 1928 Qantas entered into a unique venture with the Reverend John Flynn of the Australian Inland Mission, when it commenced operation of the world's first permanent Flying Doctor Service from its base at Cloncurry. Although John Flynn had conceived the idea of such a service several years before — and Qantas had been able to supply an aircraft suitable for the job since 1924 (the DH50) — the operation could not commence until an inexpensive but reliable means of radio communication was developed. Then in 1927, Alf Traeger perfected his ingenious foot-pedal-powered wireless transmitter, which could link remote homesteads with the aerial ambulance centre at Cloncurry.

The first and only imported Qantas DH50 (G-AUER, originally named *Hermes*) was renamed *Victory*, and the aircraft's registration was changed to VH-UER. (In 1928 the International Commission for Air Navigation revised its system of using one letter to identify the nationality of an aircraft; the countries of the British Empire had used the letter 'G' since 1921, but Australia was now allocated the letters 'VH' followed by a hyphen and a further three letters.) Under Arthur Baird's direction, Bob Struthers and George Dousha built a special stretcher to fit inside the DH50. They also modified the cabin, cutting out two seats so that the stretcher could be secured, and changing the position of the remaining two seats to allow the doctor and nurse easy access to the patient while in flight.

Under the agreement with the Australian Inland Mission, Qantas was to supply an air ambulance aircraft on demand and a pilot to fly it, based at Cloncurry. The first Qantas pilot permanently attached to the Flying Doctor Service was Arthur Affleck (Eric Donaldson would fly the service from 1931 to 1937), and the first flying doctor was Dr St. Vincent

Welch. When additional bases for the Flying Doctor Service were established in the mid-1930s at Normanton and Charleville, services from these centres were also operated by Qantas. For twenty-one years Qantas and the Australian Inland Mission cast a mantle of safety over the remote areas of western Queensland, operating the Flying Doctor Service without causing any injury to a doctor or patient — an incredible achievement, considering the type of bush flying involved. In April 1949 the Flying Doctor Service was taken over by Trans-Australia Airlines.

On 7 February 1925 Lester Brain had opened the first extension of the original Charleville–Cloncurry route, flying the newly repaired DH4 from Cloncurry to Camooweal via Mount Isa. Frank McNally was his engineer on that flight. A second route extension began operation on 1 July 1927, when Hudson Fysh flew the Qantas-built DH50A *Perseus* from Cloncurry to the isolated township of Normanton on the Gulf of Carpentaria. Then, on 17 April 1929, two DH61 aircraft were brought into service for the inauguration of the Charleville–Brisbane route extension. At last Qantas had secured its long-awaited objective of a capital city connection.

The DH61 was a 480 hp biplane with a cabin that accommodated eight passengers as well as the pilot and offered the latest in aircraft luxury — it was the first aeroplane to be fitted with an in-flight lavatory. Unfortunately, the introduction of the two DH61s (VH-UJB, named *Apollo*, and VH-UJC, named *Diana*) was not without problems. It soon became apparent that the aircraft's Bristol Jupiter 480 hp Mark XI engine was nowhere near as reliable as the excellent Mark VI version on the DH50Js, suffering constantly from chronic over-oiling of plugs, collapse of pistons, seizures and excessive oil temperatures. As a result the DH61s were involved in frequent forced landings, which sometimes required extensive and complicated repairs in the field before the aircraft could continue its journey.

Despite the fact that Bristol, the engine's manufacturer, had no legal responsibility, they lent Qantas a Mark VI engine to carry on with and rushed an expert from England to investigate. His name was Johnny Walker, and under his direction the two Mark XI engines were rebuilt to 'as-new' condition, but they continued to fail, causing the Civil Aviation authorities to threaten withdrawal of the DH61s. So serious was the problem that Qantas were only just able to keep the new Brisbane–Charleville service operating.

Qantas completed its first million miles of flying in June 1930, coinciding with the transfer of the company's head office from Longreach to the Wool Exchange Building in Brisbane. However, the main engineering work was to remain at Longreach until 1931.

Aircraft built by Qantas at Longreach between 1926 and 1929

Type	Date of acquisition	Name	Registration	Engine type
DH50A	18 August 1926	Iris	G-AUFA	240 hp Siddeley Puma
DH9C	5 February 1927	Ion	G-AUFM	240 hp Siddeley Puma
DH50A	11 June 1927	Perseus	G-AUFW	240 hp Siddeley Puma
DH50A	2 August 1927	Pegasus	G-AUGD	240 hp Siddeley Puma
DH50J	30 March 1928	Atalanta	G-AUHE	450 hp Bristol Jupiter Mark VI
DH50J	12 May 1928	Hermes	G-AUHI	450 hp Bristol Jupiter Mark VI
DH50A	22 January 1929	unnamed	VH-UJS	240 hp Siddeley Puma
DH50J	3 August 1929	Hippomenes	VH-ULG	450 hp Bristol Jupiter Mark VI

CHAPTER

3

PLANS OF EMPIRE

Arthur Baird had long since gained his 'B' commercial pilot's licence when he and Hudson Fysh became permanently based in Brisbane during 1930. Both were shown on the company's records as 'Reserve Pilots'. The chief Qantas pilot at this time was Lester Brain, but as manager of the Brisbane operations in charge of sales, demonstrations, agency tours and taxi trips, he too was classed as a 'reserve'. Russell Tapp was given the Brisbane–Longreach route, Eric Donaldson flew Longreach–Camooweal, and Charles Scott was assigned the Cloncurry–Normanton extension. The company's licenced ground engineers, as at 31 December 1929, were listed as follows:

Baird W.A.	Longreach (A,B,C,D)
Beadle H.J.	Brisbane (B,C,D)
Brain L.J.	Brisbane (D)

Compston P.H.	Longreach (B,C,D)
Hagarty N.D.	Longreach (A,C)
Jones R.T.	Longreach (B,C,D)
Williams A.H.	Longreach (B,D)
Wright D.H.	Cloncurry (B,C,D)

The classifications were:

'A' class	Inspection of aircraft (airframe) undergoing complete overhaul.
'B' class	Inspection of aircraft Engines undergoing complete overhaul.
'C' class	Inspection of aircraft (airframe) before flight.
'D' class	Installation and inspection of aircraft engines before flight.

There were also special 'X' class licence categories, which included welding, machining, line boring, radio, electrics, instruments, and so on.

The inclusion of both Lester Brain and Neville Hagarty in the list is interesting, particularly as Lester Brain was essentially a pilot and Neville Hagarty was employed at Longreach in 1926 as a woodworker. Another curiosity in the original document from which the Qantas ground engineers named above were taken, was Herbert Avery (proprietor of the Western Motor Works at Longreach and father of the first Qantas apprentice Jack Avery); although not shown as a Qantas employee, he was nonetheless listed as being the holder of both a 'C' and 'D' ground engineer's licence.

The Qantas fleet now consisted of two DH61 Giant Moths, two DH50Js, four DH50As, one DH60 Moth (G-AUFL, named *Swift*) and two DH80 Puss Moths (VH-UPA and VH-UPQ, both unnamed). The DH80 Puss Moth carried two passengers plus the pilot and was powered by a 120 hp Gipsy III engine. Two were bought to replace the Qantas-built DH50As, three of which were sold by the end of 1930. *Iris*, the first of the Qantas-built DH50As, had been modified and was now being used to supplement *Victory* as a Flying Doctor Service aircraft and would continue to function as a reserve aerial ambulance until it too was sold in 1936.

Despite the fact that problems with the Bristol Jupiter Mark XI engines on the two DH61s continued to plague the Qantas engineers (there were eighteen forced landings in 1930 as a direct result of engine failure, well over twice the number of 1929 and more than any other single year of operation to date), *Diana* and *Apollo* continued to fly as the company's flagships on the Brisbane–Camooweal route until they were superseded by the Commonwealth-class DH86 airliners in 1935.

Meanwhile, the Civil Aviation Authority had decided to abandon Eagle Farm as Brisbane's main aerodrome because of the expense involved in building much-needed runway extensions over swampy ground, and in January 1931 Qantas moved its Brisbane operation. The flying school's small hangar at Eagle Farm was dismantled and reconstructed next to the company's new hangar at Archerfield. Archerfield was farther from the city and to the south, but it proved to be a very good airport and served Qantas well for the next fifteen years.

It now became appropriate for Qantas to consolidate its base of operations at Archerfield, and so the main engineering functions were finally transferred from Longreach. However, the historic first company hangar would remain for many years as a maintenance and repair depot for Qantas and other operators, including a growing number of private aircraft owners.

On 25 April 1931 Qantas DH61 *Apollo* left Archerfield carrying the first experimental air mail to England. This was to be the shape of things to come. With Russell Tapp at the controls, the flight went only as far as Darwin, for Qantas did not have a multi-engined aircraft to fly the over-water sector to Koepang (on the island of Timor). At Darwin the mail was handed to Charles Kingsford-Smith and G.U. 'Scotty' Allan, who then flew to Karachi in the *Southern Cross*, for transfer to Britain's Imperial Airways service to London. The mail reached London on 14 May, nineteen days after leaving Brisbane. Despite the success of the experiment, it would be three and a half years before a regular air mail service would link Australia and England.

In February 1934 Qantas bought a DH83 Fox Moth (VH-URI, unnamed) to replace its original Flying Doctor aircraft *Victory* which, after almost ten years of unfailing service to the company, was about to be sold. Although smaller than the DH50, the Fox Moth was less expensive to operate, more reliable and easier to get in and out of rough terrain. It was powered by a single 120 hp Gipsy Major engine and could carry up to three passengers and the pilot. With its cabin interior modified to allow for the accommodation of a patient on a stretcher, the Qantas Fox Moth became the subject of a 1934 Cinesound Review newsreel about the aerial ambulance service.

It is now necessary to look at what was happening in aviation outside Australia in 1931 to understand why Qantas was forced into a decision that would result in it becoming an international airline. The Dutch airline KLM was operating the world's longest regular mail and passenger service between Amsterdam and Batavia (now Jakarta, on the Indonesian island of Java) and it was known that they wanted to extend their route to Australia. Britain's Imperial Airways flew a regular scheduled service between London and Karachi and were soon to extend to Singapore through Delhi; they also had plans to continue on into Australia without the participation of any Australian airline.

Qantas was operating a successful service along the route most likely to be taken by any new airline flying into Australia and could see that it must either participate in any such venture or be pushed off the route it had pioneered and be forced to close its doors. Fortunately for Qantas, the new Controller of Civil Aviation, Captain Edgar Johnston, and the then Minister for Defence in the Lyons' Government, Sir George Pearce, were united in the belief that an Australian airline company must run as far as Singapore, and tenders were called to operate the service.

It was obvious to Qantas that, with overseas operations involved, they were too lacking in experience and finance to go it alone. Intense negotiations ensued, which resulted in Qantas eventually entering into an equal partnership with Imperial Airways, forming a new Australian company — Qantas Empire Airways Limited — to tender for, and if successful, operate the Australia–Singapore airmail service. The company was registered in Brisbane on 18 January 1934. Arthur Baird was appointed Works Manager of Qantas Empire Airways, and Jack Avery, Bill Bennett and Eric Kydd, having completed their apprenticeships by this time, were shown on the company's staff register as 'Engineers'. (From this point on, the new company will be referred to as 'QEA' to distinguish it from the original Qantas company).

QEA tendered with five radio-equipped Commonwealth DH86 cabin biplanes, which were capable of crossing the 512 miles of open water over the Timor Sea with a high degree of safety. They were each powered by four Gipsy VI engines of 205 hp each (the development of which had been sponsored by Imperial Airways) and could achieve a cruising speed of 145 miles per hour. The DH86 was the result of a special order by QEA to De Havilland to produce an aircraft capable of performing well on the Brisbane–Singapore service. It met all the required conditions:

- It had to achieve an exceptionally high rate of climb, allowing for successful operation from boggy aerodromes, a quick climb to the cooler upper air in north Australian summer conditions, and the ability to continue its journey should one, or even two, engines fail.
- It had to have ample range for the Timor Sea crossing to be made in safety in the face of strong headwinds.
- Provision had to be made for the accommodation of a crew of two: the captain/navigator and a first officer wireless operator.
- It had to be capable of flying at night and provide modern standards of comfort for up to ten passengers, particularly good ventilation, the minimisation of noise, and specially designed adjustable chairs.

Amid stiff competition from West Australian Airways and Australian National Airways, QEA emerged the victor when Prime Minister Lyons announced on 19 April 1934 that the contract for the operation of the Brisbane–Singapore service had been awarded to Qantas Empire Airways.

The company's first DH86 (VH-USC, named *Canberra*) arrived in Brisbane on 13 October 1934, having been flown from England by Lester Brain, reportedly 'entirely free from all mechanical troubles'. A spare engine and numerous machine and engine spares were carried on the nineteen-day trip. With the commencement of the new service just two months away, QEA recruited Scotty Allan as a pilot for the Singapore route. He was far more experienced in long overseas flights than the company's other pilots and was a valuable addition to the team.

In preparation for the expansion of its business, QEA's head office was moved from the Wool Exchange Building to larger premises in Creek Street, Brisbane. Cedric Turner was appointed QEA's Chief Accountant, having been chosen for the job by Imperial Airways (he would have more influence on the emergence of Qantas as a major world airline than any other person). Bill Kirk was employed as an engineer and stationed at Cloncurry with Dudley Wright, and Norm Roberts began his long and celebrated engineering career with Qantas as a mechanic at Archerfield.

One day, early in Cedric Turner's career with QEA, he decided to make a spontaneous inspection of the company's operation at Archerfield. Having no transport at his disposal, he caught a tram to the end of the line, from where he made a telephone call to Arthur Baird at the QEA hangar. Norm Roberts was in Arthur Baird's office when the phone rang. After a brief conversation with his caller, Arthur Baird turned to Norm and said, 'What do you think of that? There's a new bloke called Turner waiting at the end of the tramline. He expected me to go and pick him up so that he could come out here and inspect the stores. I've told him there's absolutely nothing that could possibly interest him here and suggested he get on the next tram and go back to Creek Street.' (Nevertheless, remembering this incident some thirty-two years later, on the eve of his retirement as Chief Executive of Qantas, C.O. Turner, in a personal letter to Norm Roberts, recalled with great affection his early days with QEA and particularly his relationship with Arthur Baird.)

QEA had agreed that De Havilland could sell its DH86-type aircraft to Holyman Airways, the company that had won the government contract to fly the over-water route between Melbourne and Tasmania. Holyman ordered two, the first of which, *Miss Hobart*, made its inaugural crossing of Bass Strait on 1 October 1934. Then, on 19 October, less than three weeks after the service began, *Miss Hobart* disappeared without trace during the sea crossing between Launceston and Melbourne, with a loss of all nine passengers and the pilot. For QEA, this tragedy and the resultant public apprehension came at a critical time as they prepared to introduce their own DH86 service to Singapore. But worse was yet to come. On 15 November, as the Commonwealth Air Accidents Committee was meeting in public session to hear evidence on the *Miss*

Hobart disaster, news came through that QEA's second DH86 (VH-USG, unnamed) had crashed near Longreach on the last leg of its delivery flight from England, also with a total loss of life.

DH86s were immediately grounded in Australia, pending completion of investigations and recommendations of the Air Accidents Investigation Committee into the two crashes. In both cases the cause of the crash was a complete mystery. Weather conditions had obviously played no part, and it seemed reasonable to assume that, with four-engined aircraft, engine failure had not caused either crash. Two kangaroo shooters had witnessed the final moments of the QEA DH86 and were able to give an accurate report to investigators: as it flew towards them in the early sunlight, the aircraft began a slow, flat turn to the right; the turn tightened, and the speed fell off dramatically as the plane went into a flat spin, rotating rapidly and losing height. It hit the dry ground about 200 yards from their camp.

Scotty Allan and Jack Avery were at Longreach and were quickly on the scene. Lester Brain and Arthur Baird arrived from Brisbane later that afternoon. The crash site and the remains of the aircraft were analysed in minute detail. QEA's other DH86, *Canberra*, was examined on the ground and tested vigorously in the air. The Brisbane *Courier Mail* reported: 'The machine was put through abrupt manoeuvres. It was throttled back to two engines, the port and starboard engines being out alternatively, and corrections were made with the rudder and variable fin to counteract the torque occasioned by this.'

It was the variable fin post and the controls and surfaces attached to it that attracted Arthur Baird's attention. The entire fin, to which the rudder was attached, could be moved to assist directional stability of the aircraft, and Baird concluded that this fin post and fin had failed. Lester Brain and Scotty Allan believed that a change in the aircraft's load distribution, which placed the centre of gravity astern of the normal position, could also have contributed to the loss of directional stability. The fin post fittings on *Canberra* were strengthened, the rudder trim tab adjusted, and the centre of gravity monitored when the aircraft was being loaded.

On 30 November 1934 Hudson Fysh issued a general staff notice, addressed personally to each employee, regarding changes that were about to occur in the company's operation. Its content and emotive language gives an interesting insight into the philosophy of the time, and quoting from the letter received by Engineer Norm Roberts, it read:

> 'Dear Sir,
>
> Qantas is about to cease to operate and Qantas Empire Airways will take up the running in a wide expansion of this old business, which has covered fourteen years of pioneering activities.

You, to whom this communication is addressed, are either an old employee of Qantas, who is taking a position with the new Company, or you are new to both Companies.

The first question which presents itself, not only in regard to the staff personally, but in regard to the whole sphere of operations is — Can the old spirit of Qantas be preserved?

The old spirit of Qantas was, undoubtedly, that of the pioneer and of unstinted co-operation of all units to make the whole a success. As operations grew and became more wide-scattered, personal contacts decreased, but co-operation and individual effort for the good of the name continued. Honesty and straight dealing, coupled with service, have always been the principles striven for, and Qantas can 'cease fire' and afford to look back without regrets.

In Qantas Empire Airways a much larger sphere of operations is opened up and the following changes are apparent:

1. Personal contact and any suggestion of a 'one man show' is further diminished. Laid down routine becomes increasingly essential and individual responsibility of each member of the QEA staff.
2. The international nature of operations immediately invites a more close and critical comparison with other World Air Services. This places us more 'on our mettle'.
3. The increased importance of a safe and efficient Service is made apparent owing to the carriage of the International Air Mail and its urgent enclosures. Each delay of the Service will make itself felt right throughout Australia.
4. No less than three other Air Services are interdependent on the scheduled arrivals of QEA aircraft. Our certain aim must be to record the highest percentage of schedule efficiency.
5. Smartness of manner and dress, courtesy to clients and such kindred matters concerning the behaviour of the Staff will, of obvious necessity, have to receive more careful attention than in the past.

In short, the new Company sets out to retain the old spirit of Qantas, but, in addition, is determined to set a standard for Air Route operation in Australia, which will be second to none.

It is expected of every member of the staff that he, or she, does nothing which would tend to destroy this high standard which must be realised in all. Specially in the initial stages when the Organisation is shaking down will co-operation be necessary and assistance freely given to the 'Other fellow'.

The fullest confidence is expressed in the Staff and you are sincerely wished a long and happy association with the new Company.

Yours faithfully,
For Qantas Empire Airways Limited

(signed) Hudson Fysh
Managing Director

P.S.: Since writing the above, the unfortunate accident to the DH86 near Longreach has occurred, which has necessitated a great deal of extra work and re-organisation on the eve of commencing the Service. I am sure that this occurrence, the result of which will place an additional responsibility on most members of the staff, will only make each one of us all the more determined to see operations through to the satisfactory conclusion which lies ahead.'

On a more personal level, Arthur Baird's contribution to the success of the original Qantas was recognised when the Board presented him with an elaborately designed leather-bound illuminated address, which is still held in pristine condition as a Baird family heirloom. Its text leaves no doubt as to the esteem in which Arthur Baird was held by the Qantas Board:

'To

WILFRED ARTHUR BAIRD

Esquire
On the occasion of the emergence of Qantas Limited into the wider field of International Aviation, we, the past and present directors of that Company, desire to convey to you our appreciation of the sterling work performed by you since the very inception of the Company.

We wish to express too, the high personal regard in which you are held by all of us — a regard which has been engendered by our observation of your character through all the years in which we came in contact with you.

It is with great happiness we recall the fact that you were an original member of the Company, forming one of the staff of three which inaugurated operations in 1920. Joining as an Engineer, your ambition to become proficient in all branches of aviation led you to learn to fly and such was your keen application, that, in addition to becoming Chief Engineer and Works

Manager, you were recognised as one of Australia's ablest and most reliable pilots. When this Company's activities ended in 1934, you had built up a reputation for strict integrity and soundness of thought and action of which any man might be proud. We have no hesitation in ascribing to you much of the credit for the successful operations of Qantas Limited.

We trust that you will enjoy many more years of health and prosperity and that Australian Aviation may long be served by your efforts.'

The address was signed personally by Fergus McMaster, Alexander Kennedy, Hudson Fysh, Dr F.A. Hope Michod, Thomas MacLeod, G. Morgan Reade, Paul J. McGinness, Norman F. White, G.A. Taunton, Theo F. Barker, F.E. Loxton, A.N. Templeton, Fred Bode and D.C. Crombie. (The reference to Arthur Baird's 'learning to fly' was obviously in recognition of the fact that he had gained his 'B' commercial pilot's licence during the 1920s; he already held an 'A' private pilot's licence, having learnt to fly at Laverton before World War I.)

On 10 December 1934, His Royal Highness Prince Henry, the Duke of Gloucester, inaugurated the first air mail service between Australia and England at Archerfield. But with its DH86 grounded, still without a Certificate of Airworthiness, QEA had no option but to use its single-engined DH61 *Diana* and DH50J *Hippomenes* to operate the service. They were therefore able to fly only as far as Darwin, where the mail was transferred to the Imperial Airways aircraft *Arethusa*. The first Qantas overseas service did not leave Australia!

By January 1935 the QEA workshop's alterations to the DH86 rudder tab appeared to have solved the problem of directional stability to the satisfaction of Lester Brain, Scotty Allan and Hudson Fysh. The Department of Civil Aviation's pilot had also carried out independent tests on the aircraft, and he too had been satisfied. But before the grounding order could be lifted, the third QEA DH86 (VH-USF, named *Melbourne*) arrived in Darwin on its delivery flight from England, where it was found that the fin bias was defective. The aircraft was flown back to Singapore (on the instructions of the Civil Aviation authorities), where QEA service engineer George Boehm discovered that the bias gear malfunction was serious. If the gear had been used to check a major lateral movement of the aircraft, it could have caused a collapse of the fin post and the fin post unit.

Meanwhile, QEA's fourth DH86 (VH-USD, named *Brisbane*) had been shipped from England and on its arrival at the QEA workshops at Archerfield was also found to be faulty — the fin bias screw had been incorrectly located. In Brisbane the fault on VH-USD was corrected, while in Singapore a new fin post was made and fitted to VH-USF by QEA engineers. These modifications resulted in certificates of airworthiness being restored and DH86s being released to operate the service for which they had been designed.

It was decided to prefix the names of all the QEA DH86s with the initials 'R.M.A' (Royal Mail Aeroplane), and by June 1935 the company had its full complement of five of the aircraft type flying on the Singapore run (the final two being VH-USE, named R.M.A. *Sydney*, and VH-UUA, named R.M.A. *Adelaide*). In purely operational terms the DH86 service to Singapore proved highly successful. Lester Brain and his pilots mastered the new techniques for long flights over the sea, and despite continued minor problems with aircraft stability, Arthur Baird and his engineering staff kept the DH86s flying. It took three and a half days to fly the 4360 miles from Brisbane to Singapore. Additional line stations were established along the route at the Fanny Bay aerodrome at Darwin, Koepang (on Timor), Rambang (on Lombok Island), Sourabaya (in Java), Batavia (also in Java), and at Seletar aerodrome in Singapore.

In October 1935 Bill Bennett was appointed Assistant Station Engineer at Singapore. During his time in the East he acted as flight engineer on several search flights, including that for Sir Charles Kingsford-Smith, missing over Borneo. In May 1936 Eric Kydd, who had completed his apprenticeship in 1934 and worked at Archerfield on the maintenance and complete overhaul of the DH86 engines, was stationed at Sourabaya, from where he flew the Rambang–Koepang sector of the QEA route as flight engineer.

In the first twelve months of the Brisbane–Singapore service, QEA flew a total of 605,173 miles without a single forced landing, and 2,075,707 engine-miles with only one recorded case of in-flight mechanical trouble. This splendid operational record was a fine tribute to the quality of workmanship of the QEA engineers, and particularly Arthur Baird, whose rigid service program for the DH86s was strictly adhered to. Servicing of the DH86s was scheduled as follows:

> ' ... After every ten hours of flying, each machine on the Brisbane–Singapore service is subjected to a keen inspection. Bracing wires, fuselage structure, undercarriage and wheels, planes and control surfaces, aircraft instruments, engine mountings, controls, propellers, petrol and oil tanks, cowlings — all are rigorously examined and checked.
>
> Every twenty-five hours, inspection for condition is made of the batteries and electrical system, and the wheels and brakes are checked for correct operation. All wooden structure, fabric, Very's pistol and cartridges, passengers' and crews' life belts, emergency rations and containers, and control cabin equipment, are carefully examined. All controls are cleaned and

lubricated; shock absorbers and wheels are greased. Oil filters, petrol filters, magneto distributors and contact breakers are examined and cleaned. Contact breaker points and valve tappet clearances are adjusted. Carburettor jets are removed and cleaned and float chambers flushed out.

At 50 hours, sparking plugs are cleaned and tested.

Every 150 hours, propellers are removed and the shafts and hubs are examined.

At the end of 600 hours' flying, the four engines of each Commonwealth Class airliner are removed and in their place are installed four engines straight from a meticulous and careful overhaul.'

[*Empire Airways* magazine, December 1935, page 2]

The DH86 was the first aircraft flown by Qantas to have medium- and short-wave 'wireless transmission and receiving sets' installed. The equipment allowed for two-way communication during flight and was also used for 'homing'. Aircraft flying the Timor Sea maintained continuous radio contact with the ground communication stations on the Australian mainland and on Timor, and were able to stay on course by constantly checking the accuracy of bearings on Darwin and Koepang during the crossing.

With the opening of the Brisbane–Singapore airmail service, Darwin aerodrome assumed a new importance. It was now the northern 'gateway to Australia', and the Department of Civil Aviation moved quickly to upgrade its facilities for passenger handling and customs processing. A government administrative building began to take shape, and by August 1934 work had already begun on a new hangar for QEA. Passengers arriving in Darwin in the late afternoon were taken for an overnight stay in a cottage on the cliff at Vestey's meatworks, which had been leased and renovated by QEA.

Night flying became possible in mid-1935 and was a feature of the QEA winter schedule between Longreach and Cloncurry. Part of the upgrading of the aerodrome facilities along the QEA route included the installation of powerful rotary light beacons on top of the Qantas hangars at Longreach and Cloncurry and at strategic points between the two towns (at Moscow, Dagworth and McKinley stations). Both airports were also equipped with boundary lights, which marked the perimeter of the landing area, and with portable landing lights, which could be moved to show the pilot the direction of the wind and the best place to land. The DH86s were also the first aircraft to have 'blind flying' instrumentation fitted — the directional gyroscope and the artificial horizon. These instruments were gyroscopically operated and required absolute cleanliness to be serviced efficiently. Dudley Wright and Jack Avery worked on them from time to time, but neither had any expertise in the overhaul of such delicate instruments, and even if they had, there were certainly no suitable facilities or equipment to perform such a task at the QEA workshop at Archerfield. All electrical work (starters, magnetos, generators, etc.) was sent out to Elphinstones in Brisbane, and instruments such as altimeters, air-speed indicators and rate-of-climb indicators were serviced and overhauled by Ted Harper at Prouds Jewellers in Sydney.

That was the situation when George Roberts drove his T-model Ford into Archerfield aerodrome on 28 October 1936, looking for Arthur Baird. (George's brother, Norm Roberts, was no longer at Archerfield, having replaced Dudley Wright as QEA's Station Engineer at Cloncurry during 1935.) George was twenty-seven at the time, and on locating Arthur Baird, asked him for a job.

'What makes you think you could be useful to us?' Arthur Baird asked him.

'Well, Mr Baird,' replied George, 'you see that T-model Ford out there, I converted it from splash-feed lubrication to pressure-feed lubrication.'

Apparently that interested Baird. 'Did you really, son? And just how did you go about that?' George gave Arthur Baird the details of the process, and Arthur Baird gave George a job, despite the fact that it was against company policy at the time to employ two members of the same family. That's how engineers were employed in 1936!

CHAPTER 4

DAYS AT ARCHERFIELD

Archerfield aerodrome was flat and exposed and, at 6 am on a July morning in 1936, it was very cold. Inside Hangar 5, the larger of the two QEA hangars, DH86 R.M.A. *Sydney* was being made ready for its long flight to Singapore. Works Foreman Dudley Wright got behind one trailing edge, apprentice Jack Drinkall moved into position behind the other and, with old Sims the caretaker steering the tail wheel trolley, they began to push the aircraft out of the hangar. Sims had only one leg, so he held the trolley with one hand and his crutch with the other. Out on the tarmac the aircraft was placed in position and chocked. Dudley Wright started all four Gipsy VI engines and allowed them to warm up before running them up in turn, checking the r.p.m. drop on each magneto. The engines were then idled for a short time before being shut down.

He had just signed the daily safety check sheet when the extended-chassis Hudson arrived from the city with the passengers. There were five this morning, and their bags were loaded into the rear locker. Young Bill Neilson, from the Creek Street office, checked the passengers on board. Lester Brain was in command today, with Len Grey his First Officer/Radio Operator. They looked smart, if somewhat underdressed in the cold conditions at Archerfield, in their khaki shorts, safari jackets and long socks. The engineering staff stood by, with Jack Drinkall manning the fire extinguishers, as VH-USE taxied to the end of the grass strip and lined up for take-off. Captain Brain gunned the engines, and another QEA flight to Singapore was underway.

Back in the hangar, after a cup of coffee, Jack Drinkall returned to work carrying out routine 25-hour maintenance on R.M.A. *Brisbane*, which had come in the day before. He worked under the supervision of Harry Beadle, the ground engineer responsible for the maintenance of the starboard inboard and outboard Gipsy VI engines on the DH86s. Besides the normal oil change and cleaning of filters, it was found that one cylinder was down on compression and had to be changed, and the exhaust manifold had cracked and required welding.

Such was a typical start to the day's activities for the QEA engineers at Archerfield. Arthur Baird was Works Manager and would begin his day by calling in to his office in Creek Street, Brisbane, before going out to the workshop at Archerfield. His secretary was Ruth Trickett (who would later become Mrs Eric Kydd), and one of her duties was to type the logbook entries of engineering work done, and have the engineer concerned sign the entry. Ruth was the only female employee at Archerfield, and she shared an office in the hangar with Works Foreman Dudley Wright.

Henry Williams was foreman of the engine overhaul shop, Bill Stone worked as airframe foreman, and Mark Webber (who would in time become chief engineer of Ansett Airlines) was one of several proficient engineers who overhauled the Gipsy VI engines. Although Jack Avery had no specific title, he specialised in several areas. At a time when the process of welding was in its infancy, Jack became an expert in the art. It was said that he could weld 'the impossible', a reference to the electron tubing from which the DH86 chairs were made. (Electron tubing contains a high degree of magnesium and, for obvious reasons, it does not react well with intense heat from an open flame.) Jack developed a special technique for welding this material, so that broken chairs were able to be repaired and placed back in service without delay.

On one occasion, a QEA DH86 was in Sydney and a De Havilland engineer noticed that one of the chairs had been repaired by the welding process. He immediately telephoned Ralph Brown (who was storeman at Archerfield at the time) and asked how the chairs were being welded. 'I can't tell you that,' replied Brown, 'but if you'd like to send all your broken chairs up here to us, we'll send them back welded.' Apparently De Havilland did just that, and the secret of the welding process remained with Jack Avery.

Welding was just one of the many areas where Jack Avery excelled. Each new type of engine that came to QEA for servicing or overhaul required specialist tools, which had to be designed and made in the Archerfield workshop. The design of these tools was Jack's responsibility, but it seems that his background (growing up in the inland during the depression years) would not allow him to plan a tool to fulfil one specific function only; each tool that he developed was multi-purpose. While this may have seemed efficient economically, it was apparently not altogether successful in practice because it usually meant that the tool did not perform any function with the adroitness it would have, had it been designed with just one job in mind.

George Roberts' forte was in the field of instrumentation and electrics. Before joining QEA, George had carried out all the electrical repairs on his family's cars and had developed an expertise in the field that no-one else at QEA had. As mentioned previously, all instrument and electrical work was being sent to outside organisations for repair when George was first employed at Archerfield, but within two weeks he had convinced Arthur Baird that he could undertake all necessary repairs quickly and efficiently in the QEA workshop, at great savings in time and cost to the company. George set up his instrument and electrical overhaul section and began his one-man operation working on the troublesome magnetos on the DH86s, which were having to be serviced after each flight to Singapore and back. This was way out of proportion to the time they should have remained in effective operation between servicing, and under George Roberts' personal care the magnetos began to last the same period as the aircraft's engine before needing to be serviced — that is, 600 hours.

The gyroscopically operated instruments on the DH86s (the artificial horizons and the directional gyros) were also becoming unstable and failing. Since cleanliness is the essential ingredient for efficient servicing and overhaul of gyroscopically controlled instruments, and there were no means in the open workshop conditions at Archerfield to obtain the cleanliness required, George constructed a special box with a Perspex lid to enable him to see inside. He cut two hand holes in the side, to which surgical rubber gloves were fitted. At one end he installed a filter and at the other end a hose connected to an air compressor, which drew clean filtered air through the box. Although a crude device compared to the million-dollar 'clean room' facilities available to instrument engineers today, the basic principle was the same, and George was able to service the DH86 instruments and achieve a degree of efficiency previously unknown to QEA.

Qantas engineers have always been known for their ingenuity, and of the many anecdotes related to me illustrating this point, none appeals to me more than this one. It seems that the DH86 service from Singapore usually arrived at Archerfield when on schedule at 11:45 am. In the hot summer months the passengers were always ushered quickly from the aircraft and the cabin door shut immediately. No doubt the passengers wondered about the reason for the hurried exit and the frantic efforts of engineers to seal the aircraft. The cause of this strange behaviour became apparent at 12 noon when the engineers took their lunch break: having just made a rapid descent from the cooler air of the upper atmosphere, the aircraft's cabin temperature remained quite comfortable, and several of the engineers, seeking relief from the heat, would scamper into the still-cool DH86 cabin to eat their sandwiches.

Not all the engineering stories from this period originated in Archerfield. Out along the line, George Roberts' brother Norm was Station Engineer at Cloncurry, and on occasions he would accompany the weekly DH50 service to Normanton (on the Gulf of Carpentaria) as flight engineer. If there were no paying passengers for the return flight, it was Norm Roberts' job to buy fresh fish in Normanton, which he then had to sell in Cloncurry to help pay for the cost of the non-passenger flight. Engineers had to be versatile in the Qantas of the 1930s.

In February 1936 the Australian Government gave approval for QEA to fly a second DH86 service to Singapore, and in May the same year R.M.A. *Canberra* inaugurated the duplicate service. With two aircraft flying simultaneously, but in different directions, between Brisbane and Singapore, the workload of the engineers in the maintenanance of DH86s virtually doubled. Despite this the QEA workshop continued to service and overhaul a number of aircraft types for outside organisations and private individuals. One such aircraft was the *Percival Gull*, which belonged to station owner John Collins. He would fly from his 'Nindooimbah Station' property southwest of Beaudesert, to Archerfield and leave the aircraft with the QEA engineers for servicing while he and his wife spent the day shopping in Brisbane. In 1937 the aircraft's Gipsy VI Mark 2 engine was completely overhauled by Jack Drinkall in preparation for the first Brisbane to Adelaide air race. The *Percival Gull* took first place in its class, piloted by Arthur Baird and John Collins.

One morning Arthur Baird arrived at Archerfield as usual in his 1934 silver-painted Dodge utility, driving in across the cattle grid and around to the back of the hangars, where he parked. In the back of the ute was a metal box and crankshaft.

'Get someone to help you bring that stuff into the shop, son,' he called to Jack Drinkall. The box contained a bundle of flying wires, sheet metal brackets, 2BA bolts of varying lengths, and a logbook. Jack retained the logbook and the crankshaft, passing the rest on to the aircraftsmen (woodworkers). The logbook identified the crankshaft as having come from a Gipsy engine Mark I, Engine No. 120, formerly the property of Bert Hinkler of Bundaberg. Arthur Baird explained, 'We're going to build a Gipsy Moth, son. See Ralph Brown in the store and find out what we've got. We have to find an engine.'

All that could be found in the store was a bit from a Mark I engine and not much else of use. However, two aero club Moths had collided in mid-air a few weeks previously, burying themselves in the ground up in the Blunder, a swampy area just south of Archerfield, and Jack figured that maybe there was something there that could be salvaged. Arthur Baird made a phone call, and Doc Sims took the QEA truck down to the aero club, returning with what was left of the two engines on board.

The remains of the three engines were stripped and cleaned, and Jack was able to obtain sufficient parts to build a completely serviceable engine. By this time Bill Heaton and Ivan Hewitt had completed the building of the fuselage and wings, and Jack was told, 'Put the engine in, son.' (The wooden Gipsy Moth did not have a tubular engine mount frame or rubber mounts — the engine had to be bolted directly onto the longerons.) The engine was placed in position, and a string line was run from the crankcase through to the fin. The nose cowl was held in place with G-clamps, the propeller was fitted, and the engine moved fore and aft until Jack's fingers (to the second joint) indicated sufficient clearance between the propeller and the nose cowl. The locating holes through the longerons were drilled and the engine attached with 5/16" BSF bolts. Although very few aircraft before World War II were finished with anything other than silver dope, the QEA-built Gipsy Moth (which carried the registration VH-ULC) was painted light blue with white wings before being handed over to Harry Mills, who operated under the name of Queensland Air University at Archerfield.

An interesting follow-up to this story occurred in 1974. Jack Drinkall had been a flight engineer for Qantas since 1954 and was preparing to fly his final sector home before retiring. He was having breakfast in the garden of the hotel at Kuala Lumpur with Captain Howard Morris and was reminiscing about his early days at Archerfield.

'Did you say VH-ULC, Jack?' asked Captain Morris.

'That's right,' replied Jack. 'It was a Gipsy Moth, and Qantas built it for Harry Mills to use at his Queensland Air University.'

Captain Morris mused for a moment before saying, 'I'm glad you're retiring, Jack. Fancy me having to admit that my Flight Engineer built the plane my father taught me to fly in!'

Archerfield aerodrome was usually a pleasant place to work, but like most open areas it experienced extremes of hot and cold. One afternoon in late December 1937, the mercury soared to 111°F (44°C). A strong wind was blowing from the northwest directly into the QEA hangar, but a passage of air (albeit hot air) was created when the large rear doors of the hangar were opened to reduce the stifling heat in the all-steel structure and adjoining workshops. All metal objects were hot to touch. Such was the setting as the following incident, related by QEA's Instrument Engineer George Roberts, occurred:

'In the southeastern corner of the hangar, DH86 R.M.A. *Brisbane* was being serviced. Each undercarriage leg was supported on the twin-arms of mobile jacks since both port and starboard landing wheels had been removed. A wooden trestle was located about 2 meters forward of the port inboard engine, providing access for fuselage and cockpit window servicing. Flight Officer Fred Stevens was in the cockpit carrying out repairs to the radio. Apprentice Lance (Titch) Loney was engaged in cleaning the port inboard engine (which had had one of its magnetos removed for servicing) and engineer Ken Butler was detailed to check cylinder compression on all four engines. Within close proximity to the rear of the lower port mainplane, senior aircraftsman Bill Stone was working on the wing section from another aircraft.

Ken Butler was going about his job methodically and entered the cockpit to ensure that all engine ignition switches were in the 'off' position. He spoke briefly to Fred Stevens and pushed all four throttles fully forward before leaving the aircraft. Moving firstly to the starboard outboard engine, he pulled the propeller through its six compressions and then repeated the procedure on the starboard inboard engine before ducking under the nose to check the port inboard engine.

The Gipsy VI engine was not supposed to start on full throttle but on this one occasion assisted by the outside ambience, it coughed once and then sprang to life, roaring at full-bore on one magneto.

Ken jumped clear quickly, but the aircraft (which was supported on the mobile jacks — each having tricycle undergear) moved forward immediately, the metal propeller striking the wooden trestle, demolishing it and casting the residue through the fabric side of the fuselage.

'Titch' Loney kept his cool and tried desperately to short-circuit the port-side ignition by placing his arms across the six spark plugs, but was thrown backwards by the force of the electric shock, sustaining severe burns to both arms. Bill Stone ran forward from his workplace and attempted unsuccessfully to clamber over the trailing edge of the wing to reach the engine controls, while the aircraft, now turning in a radius toward the store, was gathering speed.

The commotion brought Arthur Baird and Ruth Trickett from their offices at the front northwestern corner of the hangar and just when catastrophe seemed imminent, the engine revs dropped, the noise subdued and the thrashing airscrew stopped. Fred Stevens, having extricated himself from the cramped position where he had been working on the radio located in the bulkhead, had closed down the throttle.

The direct cause of the near-disaster was a break in the magneto earthing lead inside the insulation which could not be seen merely on visual inspection — an indirect cause was the unusually high temperature of the day. When the throttle was opened, the interconnected magneto control caused the lead to stretch thus parting the broken wire ends. Closing the throttle reversed the control and contact was regained.'

In August 1935 Arthur Baird went to England for six weeks to study the engineering side of the Imperial Airways operation from Singapore to London and to visit the 'more important' aircraft factories in Britain. On his return to Brisbane he reported that he had seen a new and interesting aircraft design nearing completion in one of the factories:

> 'The features of construction are extremely clever, though straight-forward and simple. Every requirement has received detailed attention and the unit when completed will be aerodynamically 'cleaner' than anything yet attempted in metal construction. It will be specially equipped for the stages and distances required for the Empire route, and will provide accommodation and space for passengers to a degree of comfort not yet attempted in an aircraft anywhere. It will be complete with all navigation instruments and the latest wireless equipment for day and night flying. Sleeping berths are provided for long-distance travellers, and the upper deck has space to provide for a Purser and for the sorting of mails in flight. Four of the Bristol Company's latest engines will be the power units, and I have yet to be convinced that there is any other aero engines in their class in the world which are superior.'

By March 1937 Imperial Airways had introduced the Short S23 C-class flying boat on their Southampton–Alexandria route, and in June the service was extended to Durban in South Africa. The first flying boat service to Karachi left Southampton on 3 October 1937, but it was not until July 1938 that the Australian Government passed the Empire Air Services Act No. 13, which allowed for the introduction of the Short flying boats to operate the Empire Air Mail Service from Australia.

In anticipation of this, a contingent of seven QEA captains, along with engineers Arthur Baird, Dudley Wright, Eric Kydd and Henry Williams, was sent to England in late 1937 to observe handling and servicing technicalities of the new aircraft type. While in England, Arthur Baird recruited Ern Aldis, an Australian who had served a five-year apprenticeship with Charles Kingsford-Smith's Australian National Airlines before joining Imperial Airways at Durban in 1934. Having completed his tour of duty with Imperial Airways in Africa, Ern was transferred to Hythe in the south of England to gain experience and licences on the flying boats. At the time he first encountered Arthur Baird, he was working as an inspector, examining the new aircraft as they were delivered from Short Bros. It was a meeting that was to dramatically alter his career path and eventually see him appointed to the position of Maintenance Director of Qantas after thirty-four years' service to the company as an engineer. Ern was working behind the engine on the wing of one of the flying boats out on the water when Arthur Baird came aboard. He told Ern that he was looking for Australian engineers with licences on the flying boats and asked if he would like to come back home and work with QEA on the new aircraft. Ern accepted without too much persuasion and arrived in Brisbane by steamer to take up his position as a ground engineer in March 1938.

Short Bros produced a total of thirty C-class flying boats, an initial six of which were allocated for operation by QEA (VH-ABA *Carpentaria*, VH-ABB *Coolangatta*, VH-ABC *Coogee*, VH-ABD *Corio*, VH-ABE *Coorong* and VH-ABF *Cooee*).They were powered by four 920 hp Bristol Pegasus 10C engines and, as Arthur Baird had observed during his visit to Short Bros in 1935, they offered the ultimate in passenger comfort and cruised at a speed of 165 miles per hour.

Arthur Baird, Dudley Wright and Henry Williams had completed their flying boat training in England by March 1938 and returned to Brisbane. Eric Kydd stayed on to accompany the first QEA flying boat as flight engineer on its delivery flight to Australia. *Coolangatta* left Southampton on 18 March 1938 under the control of Captain Scotty Allan. Eric Kydd was joined on the flight by another new recruit, G.J. (Jan) Aldous, who had worked as an engineer with the Percival Aircraft Company and with De Havilland as senior inspector at their London Experimental Division before joining QEA in 1937.

Meanwhile back at Brisbane, QEA's own training program for the flying boats was well advanced. Lester Brain bought a 12 foot dinghy for pilot training, and despite Arthur Baird's reservations the company acquired a small Saro A17 amphibian, Cutty Sark, from Keith Caldwell for £700 ($1400) to use as a trainer on the Brisbane River. Arthur Baird thought its hull was unsound and the craft too small for training QEA riggers for water work. However, he need not have been too concerned about the craft's inadequacies, for the Cutty Sark came to grief when an attempted landing was made on the Brisbane River with its wheels down.

Coolangatta completed its delivery flight, arriving in Brisbane on 2 April 1938, sixteen days after leaving Southampton. But it would be four months before flying boats could begin operation on the Empire Air Mail Service — four months that would see the birth of a bold and enterprising era for QEA, as the company relocated its base of operation to Sydney.

CHAPTER

5

SPANNERS IN THE HARBOUR

The first experience with an Empire C-class flying boat for most of the engineering personnel based at Archerfield was when Imperial Airways *Centaurus* (G-AUDT), under the command of Captain John Burgess, arrived on a survey flight from the United Kingdom on 6 December 1937 and moored on the Brisbane River. The aircraft, being water-based and of all-metal construction, presented a whole series of new challenges to the QEA engineers, not the least of which was the automatic pilot. It was a model A2 produced by the Sperry Gyroscope Company of Long Island, New York. Unfortunately for *Centaurus'* crew, the automatic pilot had become unserviceable within a few hours of departure from Southampton, which meant the aircraft had to be flown manually for the duration of the flight. Although QEA was completely devoid of any technical information or experience with this new and complex instrument, Arthur Baird, when told of the problem, was convinced he had the expertise at his disposal to rectify the fault. He handed the job to George Roberts with the brief remark, 'See what you can do with the thing, son!'

George found that the ball bearings supporting the rapidly spinning gyroscope had failed. A quick trip to the local bicycle shop, where suitable replacement ball bearings were found, a spark of common sense, and some precision handling using the engineering workshop's machine tool facilities, and the automatic pilot was back in operation. Centaurus remained under its control for the remainder of its survey flight and eventual return to the United Kingdom. Interestingly, it was the first time in the history of commercial aviation in Australia that an automatic pilot controlled an aircraft through all axes of flight.

On the weekend of 28 May 1938 QEA moved its Head Office from Brisbane to Shell House in Sydney. Then, on 10 June, Lester Brain and Arthur Baird flew to Sydney aboard Coolangatta and Coogee, alighting at Rose Bay on Sydney Harbour, the site chosen for the QEA Flying Boat Base. Among the QEA engineers who transfered from Archerfield to Rose Bay were Dudley Wright, George Williams, Harry Beadle, Earl Biddell, Henry Williams, Norm Roberts, George Roberts, Norm King, Phil Compston, Noel Miller, Ted Cloherty, Bill Stone, Ern Aldis, Jan Aldous and Neville Hagarty. They all boarded at a rooming house at 10 Balfour Road, Rose Bay, close to where the base was to be set up.

Not everyone was relocated to Sydney, however. The 1628 mile Qantas internal route from Brisbane to Daly Waters was to continue operating using the company's fleet of DH86s, and an appropriate number of flying staff, ground engineers and traffic personnel were therefore left intact at Archerfield. Jack Avery remained in Brisbane and became Works Foreman when Dudley Wright moved to Sydney. Jack Drinkall, George Ditchman, Bill Heaton, Ivan Hewitt, Oliver (Olly) Shelley, Bruce Caswell, Ian McDonald, Lance (Titch) Loney, Doug Chambers and Ernie Chambers also stayed on, and Ralph Brown was made Office Manager (whilst still retaining his responsibility for stores), with Ruth Trickett as his secretary.

Rose Bay was in a lamentable state of unpreparedness for the operation of the flying boat service, due to start in August. A terminal building for passengers had been constructed, and piles that would eventually support a slipway could be seen in the water. But with no slipway, the flying boats could not be beached, so engine maintenance and repairs had to be carried out on the water for the more than twelve months it would take the Commonwealth Government — plagued by political argument about the Rose Bay location — to complete construction of the promised slipway and hangar facilities.

In the meantime, emergency measures had to be taken. Two Nissen corrugated-iron huts were built on the harbour's edge, one as an engineering and instrument workshop and the other for the catering section. At the rear of the engineering hut Arthur Baird had a small office, outside which were two typists' desks where the first two female employees at Rose Bay worked. One was Arthur Baird's secretary, Gladys Lewis (who would later become Mrs Bert Ritchie), and the other was Ruth Timmins. Several wooden benches were set up in the open area remaining in the hut, where anything that was removed from the aircraft — from the most delicate instrument to a complete Bristol Jupiter engine — had to be overhauled and repaired. A small, disused tin shed behind the Rose Bay swimming baths was leased from Woollahra Council and used as a store, under the control of Ted Cloherty.

Because engine maintenance and repairs on the flying boats had to be carried out while the aircraft was moored on the water, not only at Rose Bay but also at most of the line stations along the route to England, Short Bros (the aircraft's manufacturer) had designed a special staging that was hinged into the leading edge of the wings, on both sides of all four engines. When lowered, it became a flat surface, which gave the engineers ready access to the upper sections of each engine. Small tubular steel ladders, each with three steps, were hung from both ends of this staging and a builder's plank positioned between them. The engineers would sit on the plank with their legs dangling over the side as they worked on the lower part of the engine.

Some of the line stations did not have this 'elaborate' lower plank set-up, so the engineers worked on the engines from a bosun's chair, suspended by a rope hanging from the propeller. This was an awkward enough position in which to work when conditions on the water were calm, but when there was a swell running and the aircraft rolled about at its moorings, it became extremely difficult — and quite expensive. The company provided only the tools that were used to perform specialised functions on the engines (such as spark plug spanners), leaving the engineers to furnish their own tool kits containing standard spanners, hammers, pliers, screwdrivers, and the like. It was not unusual for an engineer to drop one of his tools into the water while working from a swaying platform under a flying boat engine, nor was it unusual for him to drop one of the company's special tools or an important engine part. The result was the same: the tool or part was lost forever and had to be replaced.

Changing engines and propellers while out on the water was also an interesting experience. The flying boats came equipped with a jib structure mounted on the engine nacelles and a hand-operated winch secured to the upper surface of the wing. It took the weight of the engine as it was unbolted and removed from its nacelle, lowering it down onto a barge, which was positioned under the wing and moored between the hull and the float. In emergency situations during the early days at Rose Bay, complete engine overhauls were carried out on workbenches in the corrugated-iron engineering hut.

Seven high-speed refuelling launches were designed, constructed and fitted out in Sydney for the Shell Company and stationed along the flying boat route at Darwin, Groote Eylandt, Karumba, Townsville, Gladstone, Brisbane and Sydney. The launches were encircled by special fenders to prevent damage to the flying boat in case of a collision. They had a fuel capacity of 1050 gallons and were capable of refuelling the flying boats at the rate of 80 gallons per minute. Fifty gallons of lubricating oil were also carried on board, and special rotary-type pumps (powered by the launch's diesel engine) discharged the oil into the aircraft. They were manned by a crew of two and fitted with a searchlight to facilitate refuelling at night.

The inaugural flying boat service from Australia under the Empire Air Mail Scheme took place on 4 August 1938, when *Camilla* left Rose Bay bound for Southampton. The agreement between QEA and Imperial Airways allowed for the flying boats to operate as an integrated fleet over the entire route, with QEA personnel flying only as far as Singapore, where Imperial Airways' crews would take over for the continued flight to England.

With the advent of the flying boats, passenger services improved dramatically. For the first time, QEA was flying an aircraft fitted with a galley for the preparation of in-flight meals, which were served by cabin staff dressed in white waiters' jackets. George Roberts recalls a meeting that was called in late 1938 to discuss methods of improving the already-exceptional in-flight catering services:

> 'Captain Lester Brain, who was Flight Superintendent at the time, was always seeking ways of providing an elevated passenger service. He considered that the breakfast menu aboard the flying boats should include, among the other foodstuffs, 'boiled eggs', so he organised a meeting at Rose Bay to discuss the matter with Works Manager Arthur Baird, Chief Steward Bill Drury and myself. My responsibilities were to include the selection and installation of the necessary cooking implements for the boiling of the eggs in-flight.
>
> The meeting analysed the project in detail and my recommendations regarding suitable equipment were judged to be in accordance with requirements. A rather delighted Lester Brain announced that he would now authorise the necessary expenditure and our course seemed set until Arthur Baird, who had contributed very little to the meeting up to this point, threw in his trump card by asking a very pertinent question, 'How do you boil an egg at ten thousand feet?'
>
> Everyone just looked at each other and nobody spoke. The project was unceremoniously scrapped and boiled eggs were never mentioned again.'

While passengers on the flying boats experienced a degree of luxury previously unknown, flying time to Singapore was no faster than the DH86s and it still took the best part of four days to get there. The route from Sydney followed the northern coastline, stopping at Brisbane and Gladstone in Queensland for refuelling before arriving at Townsville for an overnight stay. From Townsville the route veered inland and crossed the undulating cattle country of northern Queensland to the Gulf of Carpentaria, where the aircraft was refuelled at Karumba and Groote Eylandt, before proceeding on to Darwin for a second night stopover. Day three and over the Timor Sea to Koepang (on Timor), then to Bima (on the island of Soembawa) and arriving at Sourabaya (on the eastern coast of Java) just before nightfall. On the fourth day the relatively short 973 mile sector from Sourabaya to Singapore required only one stop for refuelling at Batavia (Jakarta).

Engineering line stations were set up by QEA along the route where overnight stopovers occurred: Townsville, Darwin, Sourabaya and Singapore. George Williams had gone to Townsville to establish a station on the shores of Ross Creek and was later joined by Ern Aldis, who flew from Sydney aboard *Camilla* on the inaugural flight. When George Williams was recalled to Sydney, Ern Aldis became station engineer at Townsville. His staff consisted of engineers Charles Brodie and Alan Cunningham and coxswain Alf Back, who operated the engineering launch *Nestor*.

Norm Roberts and Eric Kydd were appointed station engineers at Darwin and Sourabaya, respectively. In Sourabaya the Netherlands Indies Government had built an administration and passenger reception office at the Tandjoeng Perak flying boat base, directly opposite a jetty leading to the pontoon landing stage, and Eric Kydd was allocated a small area within the building to house his stores and spare parts. He had a team of three engineers from Australia (Ken Butler, Tom McCourt and Wally Martz) and a number of Javanese labourers employed at the base. Passengers were accommodated overnight at the Oranji Hotel. Bill Bennett had returned to Sydney during the year to undertake a flying boat handling course and familiarise himself with the Bristol Pegasus engines; he was posted back to Singapore shortly after the commencement of the service but was now in charge of the base, after his promotion to Station Engineer.

The line stations were equipped in kit form by Imperial Airways, each station being given an A, B or C classification depending on its function and the frequency of its operation. The kit included everything from spare engine cylinders, pistons, nuts, bolts and sheetmetal to mops, brooms and toilet paper, the quantities of each varying with the classification of the station.

The flying boats were scheduled to arrive at the night stopover ports in the late afternoon. Passengers were ferried ashore in a small launch and taken to their place of accommodation, while the Captain handed the aircraft's technical log to the Station Engineer. The log contained details of any defects in the aircraft's operation that had become apparent during the day's flight, all of which had to be rectified before the scheduled departure time of 6 o'clock the following morning. The Station Engineer was usually faced with an assortment of defects, which varied from problems with the engines, electrical systems, radios, cockpit instruments and hydraulics, to damage to the aircraft's metal hull. But it was the Bristol Pegasus engines that were in most constant need of attention. They were often found to have low cylinder compression, burnt exhaust valves, worn valve rocker gear and ignition defects. According to Ern Aldis, 'it was common to change three or four cylinders and pistons per week, by torchlight, suspended over the water on a plank or bosun's chair, constantly fearful of dropping expensive and critical engine parts or special tools into the muddy depths of [Townsville] harbour.'

On one memorable occasion the engineers at Townsville had just completed their maintenance on an engine and were replacing the cowling. This was a three-segment wrap-around cowl, and just as the bottom section was being placed into position, it somehow fell away and dropped into the water, sinking immediately to the bottom. It was 2 am, the aircraft was due to depart in just four hours, and they had no replacement part. The cowl was a complex design, with an air scoop at the bottom to feed air to the carburettor, and without it the flying boat was demobilised. The engineers hurried ashore to their corrugated-iron engineering shed on the shore of Ross Creek, where, working against the clock, they fashioned a substitute piece from a sheet of duralumin, nuts, bolts and some rivets. So pleased were they with the result of their ingenuity that they wrote 'Please return to Townsville' on the inside of the new piece before fitting it to the engine with little time to spare for the 6 am departure. Although the part may not have met with DCA approval, it got the aircraft to Singapore without incident, where Bill Bennett replaced it with a factory-made cowl segment. The substitute part was returned to Townsville as requested, but it was never called back into service.

When an engine had to be changed at one of the line stations, it had to be done during daylight hours because of the dangers involved in performing such a major undertaking out on the water in the dark. This invariably meant that the aircraft was delayed for twenty-four hours, making it a day late getting to the next engineering station. When this occurred, it usually coincided with the on-schedule arrival of the flying boat travelling in the opposite direction. This presented a significant challenge to the engineers who, with no additional resources, had to ensure that both the north- and south-bound aircraft were ready for early departure the next morning. With both captains wishing to be first away and urging that their flight be given priority of service, the Station Engineer had to be very diplomatic and have a high degree of determination.

It was the responsibility of the Station Engineer to provide assistance to the regional stations (not manned by engineering personnel) north and south of his base. In the case of Townsville, this responsibility extended north to Karumba and south to Gladstone. On one occasion, a northbound flying boat became stranded at Karumba because of engine failure. A relief flying boat was despatched from Sydney carrying a spare engine, and Ern Aldis and Alan Cunningham joined the flight at Townsville. The spare engine was mounted in a wooden transport stand and stowed in the rear cargo compartment of the aircraft. It was 'bare' and needed both a front mounted exhaust ring with short pipes connecting to the cylinder exhaust ports and also a 'gill ring' assembly (which controlled the airflow past the cylinders at the rear) to be fitted before the engine could be installed on the demobilised flying boat. Knowing that it would take about two hours to affix these items, the engineers locked themselves into the cargo compartment and working with lead lights and torches proceeded to build up the engine during the four-hour flight to Karumba.

Karumba is located in the remote southeast corner of the Gulf of Carpentaria at the mouth of the Norman River, and the flying boat mooring was in the middle of the river, which flowed quite quickly most of the time. The flow of water wanted to 'stream' the hull of the flying boat, but the high mounted fin and rudder surfaces were wind-reactive and would point the aircraft in the direction of the air current. As a result the wind would often align the flying boat at right angles to the water current, creating a dangerous situation for a heavily laden refuelling barge or a work launch with a spare aircraft engine on board as it approached a moored flying boat. It took a skilful coxswain to manoeuvre a launch and avoid a collision in these circumstances. (Although the flying boats were large aircraft for their time, to meet design lightness requirements the external skin was only about 1 millimetre thick and was easily punctured or dented.)

When Ern Aldis and Alan Cunningham arrived at Karumba, they found that the river flow and the wind were at right angles to each other, making the task extremely difficult. They began removing the failed engine from the nacelle using the wing-mounted derrick and jib, lowering it slowly to the surging workboat below. Then they removed the new engine from the rear cargo compartment of the relief aircraft and ferried it across the rolling swell of water to the disabled flying boat moored in the middle of the river. Such was the skill of the engineers that the engine was installed and tested successfully in the traumatic and unstable conditions on the Norman River, and the stranded passengers were able to continue their journey to London just twenty-four hours behind schedule.

In December 1938 QEA lost its first flying boat. *Coorong* had alighted at Darwin without incident and had been secured to its mooring for the night, when high winds and a heavy swell developed. Station Engineer Norm Roberts became concerned and went out to the aircraft in the launch. In an attempt to ease the strain on the mooring rope and storm pennant, he started the engines and let them run, but the *Coorong* subsequently broke loose and was blown onto the rocks of the breakwater. Damaged well beyond the resources of QEA to effect repairs, it was dismantled and sent by ship to Short Bros in England, where it was fully rebuilt and eventually re-entered service.

By the end of January 1939 the new slipway at Rose Bay had been completed and was ready for testing. It was of timber construction and connected with a concrete apron that extended to the area adjacent to the new hangar (still in the early stages of construction). The aircraft chosen to be the first ashore was Imperial Airways' flying boat *Cameronian* (G-AEUE). Special beaching gear had to be attached to the aircraft before it could be towed up the slipway, and this was floated out and fastened to the hull by the beaching crew, who had to swim with it out to the slipway mooring. On the concrete apron stood the company's brand new Caterpillar tractor, ready to haul the giant aircraft out of the water and up onto the tarmac. Large rubber blocks had been inserted between the tracks of the caterpillar to prevent damage to the concrete. With the beaching gear in place and the towing cable affixed to the rear of *Cameronian*, the tractor began to move slowly forward. The aircraft came out of the water and moved up the slipway tail first. It was almost to the top when the tractor ran into a wet patch on the concrete and began to lose traction. Immediately the weight of the aircraft took over, and *Cameronian* began to roll back down the slipway, dragging the tractor behind it. Arthur Baird, who had been directing the operation, turned to George Roberts and remarked in his usual unflappable manner, 'Keep an eye on this, George. We're going to see something interesting here in a moment.'

And see something they did! As the free-wheeling flying boat gathered momentum on its run back down the slipway completely out of control, the terrified tractor driver jumped for his life as his machine was drawn into the harbour, disappearing from view beneath the surface. The investigation of the incident that followed revealed that, although the tractor had the power to haul the flying boat up onto dry land, it did not have the weight, hence its loss of traction when it ran into the water patch on the concrete. A larger, heavier tractor was ordered, and subsequent tests on the new slipway proved successful.

With the exception of the few engines that were repaired as an emergency measure in the Rose Bay engineering hut, all Bristol Pegasus engines from the flying boats were sent back to England for overhaul. But by mid-1939 it was anticipated that the disturbances in Europe could escalate, causing a break in the air mail service to Great Britain. This concern highlighted the need for Australia to be independent in its flying boat operation.

In a move to ensure self-sufficiency in the event of war breaking out, QEA let a contract for the construction of a two-storey engine overhaul workshop at Kingsford-Smith Airport at Mascot. Built at a cost of £7000 ($14,000) the building was situated in Vickers Avenue on the eastern side of the airport; it was completed and ready for occupation on 18 August 1939, just thirty-nine days after the turning of the first sod and four weeks within the contracted time. Designed by architects Rudder and Grout primarily for the overhaul of the Bristol Pegasus engines, the imposing Mascot facility also housed engineering offices and a large supply of spare parts. It also had provision for the quick take-down, inspection and reassembly of the engines, as well as a specially structured U-shaped test house for engine tests. A unique feature of the complex was the design of the arched roofing members which carried the skylights; using the 'saw-tooth' principle, they were positioned to allow the maximum amount of natural light to fall on the machines and workbenches.

The opening of the engine overhaul workshop represented a milestone of achievement for QEA. Not only did it further the company's scheme of comprehensive service and self-containment, but it was also the first facility built by Qantas for the specific purpose of overhauling aircraft engines. At this time, the QEA staff numbered 285. Arthur Baird as Works Manager controlled 114 engineering personnel, 40 of whom were employed at the new Mascot workshop, where Henry Williams was Works Foreman. The systematic overhauling of the Bristol Pegasus engines followed an interesting series of set procedures:

> 'Each engine is first put through a stripping and cleaning process where each part is dismantled and thoroughly cleaned, either by washing in special baths, sand blasting or by wire brushes which rotate at high speeds. When this job is done, the complete set of engine parts is categorically labelled and passed along to be checked and inspected for wear or faults.
>
> Each part is subjected to meticulous inspection. A specially constructed detector

is employed to discover cracks and flaws so minute that they are otherwise impossible to detect, even with a strong magnifying glass. Checking by set standards and with micrometer gauges follows — parts worn beyond a certain specified tolerance are rejected and scrapped. Corresponding new spare parts are brought from stores to take their place in the complete set.

Then comes the assembly process. The set is broken into various components, which proceed to the different assembly benches. Those parts belonging to the fuel system are assembled in the carburettor or supercharger section. Valves, pistons and gudgeon pins to the cylinder section; the magneto section covers all high tension leads, spark plugs, coils and parts relating to the ignition system; and so on.

Throughout the entire overhaul process, the larger parts are covered by clean cloths to protect them from dust or other abrasive elements that may be floating around in the air and could subsequently cause damage to the highly polished surfaces or working parts. Smaller items are sealed in screw-topped glass containers.

First to be assembled is the crankshaft and main bearing. The alignment and adjustment of the 'maneton' bolt is a job which requires extreme care and precision, for it is on this 'king' piece that the rest of the assembly is arranged. The remainder of the engine is assembled in a strict routine order after the crankshaft and casing have been attached to a rotatable assembly frame.

This frame allows the radial engine to be turned into any position and provides easy accessibility in all positions.

Two assemblers build up the engine, one working on the back and the other on the front, and the work proceeds along the various stages of pistons, cylinders, valves, blowers, reduction gear, pumps, etc., until it is ready for the final assembler. His job is to time the engine electrically and mechanically and make the necessary adjustments for 'running-in'.

The last process is actually performed in two stages, namely 'motoring-in' and 'running-in'. The engine, when finally assembled, is then 'motored-in' at the correct oil pressure and temperatures, being driven by a powerful electric motor to ensure that all working parts are operating smoothly.

Having passed all stages, including the 'motoring-in' period, the final test takes place in the specially designed wind tunnel under the engine's own power. The whole assembly is bolted onto a bed similar to that on the flying boat and connected to a duplicate set of instruments in a glass-enclosed observation room.

A heavy four-bladed test fan takes the place of the three-bladed type used on the aircraft. Similar conditions to actual flying are artificially created as the engine is run up at varying speeds for specific periods and a close check is kept on its operation. Boost pressures, revolutions per minute, individual cylinder head temperatures, oil temperatures, oil and petrol consumptions are all carefully read and checked during the full throttle period of the test run.

All engines passing this severe final test are passed out ready for service.'

[*Empire Airways* magazine, August 1940]

It is a matter of record that the Bristol Pegasus engines overhauled by QEA engineers at the Mascot facility gave a significantly improved period of service on the flying boats, resulting in fewer engine changes at the line stations along the route. With the provision of slipway facilities at Rose Bay, engine changes and servicing no longer had to be carried out over water. They did, however, still have to be done in the open, because the long-awaited hangar was not completed and ready for use until October 1939.

To assist with the movement of an engine to and from a flying boat on shore, the chassis of a 1914 model Carrier truck was purchased. A jib was built onto the chassis as a fixture, which meant that it could be lowered or raised but could not rotate. The truck therefore had to be backed up to the aircraft and the jib centred above the engine to prevent it from swinging from side to side when it was unbolted from the nacelle, and that was a relatively easy procedure. However, reinstalling an engine required a larger degree of operator skill, because the crane had to be lined up exactly so that the attachment bolts could penetrate the nacelle in the correct position. Despite its inability to rotate, the Carrier crane was an essential and efficient piece of equipment, and it continued to give good service until it was finally retired at Mascot during the mid-1950s.

By the time the Rose Bay flying boat hangar finally came into operation, war had been declared in Europe and the romantic era of peacetime flying boat services was fast coming to an end. But the QEA hangar was to play an important strategic role in assisting Australia's war needs. It was built on the harbour's edge and ran parallel to the shoreline, at ninety degrees to the slipway and concrete apron. It provided accommodation for the simultaneous servicing of two flying boats, positioned tail first, and a number of annexes furnished as engineering administrative offices and servicing workshops. Along the northern wall of the hangar were the office of the engineer in charge (Dudley Wright) and a lobby area where Arthur Baird's secretary also acted as receptionist. Above the engineer's office, at an elevated mezzanine level, was the Works Manager's office, which provided Arthur Baird with a commanding view of the entire hangar complex. Behind the lobby lay the dual-level stores installation and the amenities area, which included showers and a lunch room. The back (western) wall of the hangar had a lower roof section, under which were workbenches where Jan Aldous and Bill Stone carried out repairs and maintenance to airframes. The southern side wall had three partitioned but open sections, which housed propellers, the machine shop and a storage area for heavy items such as spare engines. Above the open storage area was the enclosed radio shop, operated by AWA under the charge of Fairley Roger De Burgh Snap — or 'Snip', as he was more commonly known.

The instrument overhaul area above the machine shop was a separate free-standing structure, which was not affixed to any part of the hangar construction. This isolation was to prevent the incidence of vibration, for it seems that the hangar moved considerably during high wind periods. George Roberts was no longer the only person in the instrument overhaul section; Ted Harper had been recruited from Prouds Jewellers, and an apprentice from Archerfield, Brian Murton, had also transferred to Rose Bay.

The Rose Bay hangar had a unique communication system, which although somewhat primitive was very effective. When Arthur Baird wished to attract someone's attention he would stand at his mezzanine office window, place two fingers in his mouth and deliver a powerful, piercing whistle, which could be heard right throughout the hangar complex. All work ceased immediately, with everyone turning their gaze to the open window where Arthur Baird stood, pointing at the individual to whom he wished to speak. The person thus identified would hurry to the Works Manager's office as work on the hangar floor returned to normal.

CHAPTER 6

ESCAPE FROM JAVA

Following the outbreak of World War II on 3 September 1939, QEA maintained a twice-weekly schedule in each direction on the Australia–England flying boat route. Although considerations of censorship prevented publication of timetable details, continuance of the service ensured that vital lines of communication between outlying units of the Empire were kept open. Then on 1 April 1940 the British Government's newly formed Airways Corporation acquired the undertakings of Imperial Airways and also those of British Airways, creating a government-owned organisation known as the British Overseas Airways Corporation (BOAC) to operate the essential land and sea plane services of the Empire.

But when Italy entered the war in June 1940 and the Empire Flying Boat route through the Mediterranean was cut, QEA's *Carpentaria* and *Cooee* began flying a 'horseshoe' service from Sydney to Cairo, then south to Durban in South Africa. The mail was taken from there to

England by ship. The change of route through Africa led to BOAC moving its headquarters, together with a large technical and administrative staff, to Durban. As the BOAC engine overhaul facilities at Southampton were now unavailable to aircraft east of the Mediterranean, the entire fleet of sixteen QEA and BOAC flying boats operating the horseshoe service became totally dependent on QEA's new Mascot workshop for overhaul of their Bristol Pegasus engines. 'That they were available', reported the *Sydney Morning Herald*, 'was a tribute to the foresight of Mr W. Hudson Fysh, Managing Director of Qantas Empire Airways Limited and the Works Manager of the Company, Mr A. Baird.' The report continued: 'Today, the Qantas engine overhaul section not only plays a large part in keeping the Empire flying boats in the air over the whole of the Durban–Sydney route, but it is also of service to the Air Force in maintaining Royal Australian Navy shipborne aircraft.'

Maintenance and overhaul facilities were eventually established by BOAC at Durban, and QEA began a program to make long-range fuel tanks for the flying boats.

At the time war was declared, two BOAC flying boats *Centaurus* (G-AUDT) and *Calypso* (G-AEUA) were in Sydney and, along with a number of other civil aircraft, were impressed by the RAAF for transport and sea patrol duties. They were converted for military use at the QEA hangar at Rose Bay under the direction of Jan Aldous. The mooring hatch door in the aircraft's nose was replaced with a bomb sight dome, and British 303 machine guns were fitted. Additional fuel tanks, designed and built in the Rose Bay hangar, were also installed to give the aircraft greater range — three positioned in the centre section of the aircraft and two in each wing.

They were repainted in heavy camouflage colours supplied by B.A.L.M. Paints Pty Ltd, and the propellers were to be black. On the first test flight at Rose Bay, which was conducted in perfect, fine-weather flying conditions, the aircraft returned with no black paint on the propellers. The representative from B.A.L.M. was called in, and he established that the paint coating had been applied to the propellers exactly in accordance with the specifications and should not have fallen off during flight. A second test flight was conducted under the same flying conditions, and the same thing happened again. This time B.A.L.M.'s Chief Chemist was brought into the investigation, and he discovered that the propellers had been washed down with mineral turps to clean off the previous paintwork, creating an unstable base for the black paint. The propeller blades were then etched and a new coat of paint applied. This time it remained intact.

The two flying boats were placed into service with No. 11 Squadron

RAAF based at Port Moresby. With highly trained QEA crews conscripted to fly them, *Centaurus* was given the new registration A18-10 and *Calypso* A18-11. Two of QEAs flying boats, *Coogee* (VH-ABC) and *Coolangatta* (VH-ABB), were also impressed for use by No. 11 Squadron in mid-1940, becoming A18-12 and A18-13, respectively. They too were converted at Rose Bay, being modified to carry American Point Five machine guns and two 500 pound bombs beneath the wings.

Like the QEA flying boats operating the horseshoe service to South Africa, the four aircraft impressed for military duties also had to undergo complete engine and airframe overhaul after a specified number of flying hours. The airframe overhauls were carried out at Rose Bay, and to assist with the task the RAAF made available to QEA some of their personnel from the Rathmines sea-plane training flight facility, which was situated on the shores of Lake Macquarie near Newcastle in New South Wales.

A18-10 was the first to be returned to Rose Bay for overhaul and was standing on its beaching gear in the hangar when Arthur Baird decided to check on the progress of the work. The floor had been completely removed, and as Baird walked up the mobile stairs and stepped into the aircraft, he came across a young RAAF serviceman sitting on a work plank smoking a cigarette. This was extremely dangerous because the aircraft was full of vapours from the fuel tanks and a devastating explosion could have ensued, so Baird reprimanded him for smoking inside the aircraft. The serviceman made no attempt to extinguish his cigarette but simply turned and replied sharply, 'What the bloody hell's wrong with you?'

The Works Manager was obviously not too impressed with this response and repeated that smoking was not permitted inside the aircraft, only to be told 'Well, mate, if it's good enough for you to smoke, it's good enough for me!' Baird, a heavy smoker, suddenly realised that he too was holding a lighted cigarette, which he had forgotten about and inadvertently brought onto the aircraft. His reaction was typical of the man; he patted the young serviceman on the shoulder and in his usual gentle manner said, 'Sorry, son. We've both learnt a lesson today, haven't we.'

When people who knew and worked with Arthur Baird speak of him today, they reminisce about his enviable personal attributes and his capacious talent as an engineer. It is well remembered that, from the very start of Qantas operations, he set about establishing a philosophy that came to be the standard of Qantas engineering. His motto, 'Near enough is not good enough', became the ethic of performance in the company's hangars and workshops for many years, and his kindly 'Do you know what you're doing, son?' endeared him to apprentices and qualified tradesmen alike.

Arthur Baird also had considerable creative expertise, and his outstanding design for the working platforms (specially constructed in the Rose Bay hangar for servicing flying boat engines) so impressed the Americans that they took the design back to their own country and began building them for their own use. The rostrums, as they were called, were built on a scaffolding principle but were mobile; caster wheels on all four corners of the structures allowed them to be located under the aircraft and locked in position to prevent movement. Their design allowed for the simultaneous servicing of both engines on each wing. They were of wooden construction, with a stairway from the ground to the main working platform, which had two workbenches (containing drawers for tools) located conveniently adjacent to each engine. The platforms were built at engine height, thus allowing engineers to work underneath and around the sides of each engine from a comfortable standing position. Work on the top section of the engine was carried out from the wing leading-edge platforms, located on both sides of all four engines. The rear of the engine — while still on the aircraft — had to be accessed through the inside of the wing. The core of the wing on the flying boats was about 5 feet deep and was entered through small doors from the aircraft's centre section. The engineers could crawl out through the wing to the rear of each engine on either side, even while the aircraft was in flight.

Fuel filters on the Bristol Pegasus engines were located one in each nacelle behind its respective engine, and the inspection and servicing of the filters was carried out with a low-voltage incandescent lamp on an extension lead providing the necessary light in the nacelle's shadowy depths. With each service, some spillage invariably occurred, the waste fuel falling onto the floor of the nacelle.

On one occasion in 1940 a flying boat was undergoing regular service in the Rose Bay hangar. The inspection and servicing of the fuel filters had been completed, but the lead light was still hooked to the ceiling of the port inboard engine nacelle. Eric Walsh was busy with periodic wing inspection and was inside the wing at the rear of the fuel tanks. George Roberts was in the cockpit servicing the instruments. Suddenly the lead light in the nacelle fell from its hook, breaking the lamp and setting fire to the vapour rising from the spilled fuel. The explosion caught George Roberts' attention, and from his observation point in the cockpit he saw Eric Walsh virtually rocket out from his precarious position inside the wing, landing on all fours on the deck of the rostrum, unhurt by the fire but suffering a few scratches nonetheless.

Then a second explosion occurred, spreading the fire to the port outboard engine nacelle and wing interior. Down on the hangar floor, Norm King seized a fire extinguisher from the hangarside fire station and ran towards the rostrum, followed closely by George Williams. As he reached the rostrum and prepared to run up the stairs with fire extinguisher at the ready, Norm King's foot missed the first step, causing him to stumble and spin around through 180 degrees. In the confusion of the moment, the trigger on the fire extinguisher was accidently released and George Williams was brought to an unceremonious dead stop by the force and volume of the expellant dousing him.

The fire was quickly quelled with the fire-fighting equipment based on the rostrum. It had been caused by a gas-vapour build-up within the wing and nacelles, and the quick action of the engineers on the rostrum resulted in no sustained damage to the engine or wing being recorded. The aircraft was allowed to depart on schedule the following day.

By the beginning of 1941 QEA staff numbers had risen to 409, and the QEA workshops at Rose Bay, Mascot and Archerfield were becoming increasingly occupied with overhaul, maintenance and servicing of aircraft for the RAAF.

When exceptionally heavy overseas mail and passenger commitments on the Empire Flying Boat Horseshoe Service began to congest air traffic between Darwin and the eastern states, QEA inaugurated a new through-service from Brisbane to Darwin via the inland route on 3 July 1941. The aircraft chosen to operate the service was a Lockheed 10A 'Electra' (VH-AEC, appropriately named *Inlander*), which was to be the company's first non-British aircraft acquisition. It operated a weekly return service to and from Darwin, stopping at Cloncurry overnight. The Lockheed 10A was powered by two 450 hp Pratt and Whitney Wasp Junior SB1 engines and could carry ten passengers and a crew of two at a cruising speed of 180 miles per hour. It had been assembled in Sydney by QEA engineers during June 1941, after its arrival by sea from its previous owner in America.

In early December 1940 Dudley Wright, one of QEA's most experienced engineers, had travelled to America to prepare for the delivery to Australia of PBY5 Catalina flying boats, which the Australian Government had ordered from the Consolidated Aircraft plant in San Diego. Because the United States had not, at this stage, become embroiled in the war, it was considered diplomatically prudent to have the delivery of the Catalinas carried out as a civilian operation. One of the conditions of purchase was that the aircraft be flown to Honolulu by American crews. There they would become the property of the Australian Government and could then be flown on to Sydney by QEA's civilian pilots, radio officers and engineers.

Dudley Wright was joined in San Diego on 1 January 1941 by Captain Lester Brain (who had been placed in charge of the delivery program), QEA's Senior Radio Officer A.S. 'Pat' Patterson, and P.G. 'Bill' Taylor, who, although not a QEA employee, was one of Australia's leading navigators and had been co-opted to direct the course of the first delivery flight. When the first Catalina left San Diego on 25 January 1941 it was only the third time in aviation history that a direct flight across the South Pacific Ocean had been attempted. The route taken was via Honolulu (where the three Americans on board left the flight), then on through Canton Island and Noumea to Sydney. The total elapsed time from San Diego was seven days, while actual flying time was 60 hours 16 minutes. It was the first aircraft to arrive in Sydney direct from a foreign port. According to Lester Brain's report of the flight, there were no mechanical troubles. 'The engine cowlings and wings were spotless on arrival in Sydney, in marked contrast to the British engines of similar type, which have four to six times the oil consumption and throw oil and grease all over the place.'

Between January and October 1941 a total of nineteen PBY5 Catalinas were flown across the Pacific without incident by QEA crews. The company engineers participating in the program were Dudley Wright, Jan Aldous, Charlie South, Charles Brodie, George Williams, Rex Mitchelhill, Ern Aldis, Colin Sigley, Norm Roberts, Bill Willcox, Ken Butler, Norm King and Neville Hagarty. With the exception of Dudley Wright and Jan Aldous, who had received their training on the aircraft type at Consolidated Aircraft in San Diego before the first and second delivery flights, all QEA engineers involved in the operation were trained on the Catalinas at Rathmines by RAAF personnel. This facility was now known as No. 1 OTU (Operations Training Unit) and had been developed as an establishment where everything connected with flying boats on active service was taught. Although QEA lost the invaluable services of Scotty Allan to the RAAF, he became Flight Officer in charge of the Rathmines base.

The company saw this as a great opportunity to add to its engineers' experience on flying boats, and even brought Norm Roberts back from Darwin and Ern Aldis from Townsville to participate in the program. All QEA engineers were given Flight Sergeant ratings for the duration of the delivery flights and at the end of the war they were afforded Returned Servicemen's classification.

In March 1939 Ruth Trickett had resigned from her position as Ralph Brown's secretary at Archerfield and had travelled to Sourabaya to marry QEA's Station Engineer Eric Kydd. Ruth was the first engineer's wife to be allowed to stay with her husband on a QEA overseas posting, but had to pay her own fare as there was no free travel for wives in those days. One of the first things she had to do on arrival in Sourabaya was learn the Javanese language. It was basically Malay, and she very soon learnt to speak it fluently, despite the fact that the local version was complicated by the addition of Dutch prefixes and suffixes. Initially Eric and Ruth lived in a suite of rooms at Sourabaya's Hotel Pension until a rented house became available. Their transport was the company's 1939 black Ford panel van.

In 1941 engineers Tom McCourt, Wally Martz and Ken Butler were moved from Sourabaya to new postings and were replaced by Dave Thomson and Jim Lamb. Both Dave and Jim were married and were given permission to take their wives Rona (Thomson) and Stavranay (Lamb) with them — Ruth Kydd having set a precedent. While the three engineers worked six days a week, including weekends, and mainly at night, their wives busied themselves working voluntarily for the Dutch Red Cross, knitting sweaters for Dutch seamen. From all accounts, it was a good life in an idyllic setting, full of fun, mischief and comradery. But unfortunately this tranquil existence was to be short-lived, as events on the world scene moved inexorably towards the drama of 7 December 1941, when Japan's surprise attack on Pearl Harbor brought war to the Pacific region.

QEA's Managing Director Hudson Fysh had arrived in Singapore on 7 December by flying boat from Sydney for a meeting with BOAC's Walter Runciman. He was asleep in his room at Raffles Hotel when, at 4:15 am the following morning, he was awakened by 'the unmistakable noise of anti-aircraft guns, bombs falling and an air raid siren wailing'. The air raids on Singapore had begun. By the time Hudson Fysh left Singapore on 14 December on his return flight to Sydney, the Japanese had begun their advance down the Malay Peninsula, and he could see that it was only a matter of time before Singapore would fall. On arrival at Sourabaya, his message to Eric Kydd was clear: 'Get the women out as soon as possible!'

A decision had been made that Singapore should be abandoned as QEA's maintenance base, and on 15 December the company's standby flying boat left Singapore with Station Engineer Bill Bennett and engineer Charlie Short on board. They took with them all the aircraft equipment and engine spares that could be carried on the flight, and set up an emergency operation at Batavia.

Despite the fact that air raids on Singapore were now being carried out during daylight hours and becoming increasingly heavy, the horseshoe route from Durban remained open, and on 6 January 1942 a shuttle service began between Batavia and Singapore, connecting with the through-service.

Ruth Kydd, Rona Thomson and Stav Lamb were all to be evacuated from Sourabaya on a QEA flying boat due to arrive from Batavia on 12 January 1942. The aircraft left Batavia on schedule with Bill Bennett on board (he was returning to Brisbane to be married), but when the aircraft arrived at Sourabaya it was found that one of the engineers' wives would have to give up her seat and wait for the next flight out. Sir Charles Vyner Brooks, the 'White Rajah of Sarawak', who was also an evacuee, had been given priority status for the seat on the aircraft. The toss of a coin decided that Stav Lamb would be the one to wait for another flight. Fortunately, she was able to be flown out the next day, but at the time the situation was so unstable that no-one could be sure if there would actually be a 'next flight out'. Ruth Kydd recalls the emotion of the moment as she boarded the aircraft bound for the safe haven of Australia: 'Imagine our feelings as we flew out of Sourabaya, leaving our husbands behind, wondering when, if ever, we were to see them again.'

The flight carried mostly evacuees from Singapore, many of them women. Ruth and Rona put themselves to work helping the flight stewards cope with the strain of their distressed passengers, many of whom were pregnant. Everyone on the aircraft had to be sworn to secrecy as the huge American fleet that was approaching Australia's shores could be clearly seen from the flying boat. On arrival in Townsville the flight was met by a large contingent from the press, eagerly seeking information on the Japanese advance. Sir Charles Vyner Brooks had foreseen the possibility of this and volunteered to take the brunt of the reporters' questioning.

QEA's Station Engineer Ern Aldis and his wife, Joan, took Ruth and Rona into their care, extending to them the hospitality of their home after the emotional ordeal of the flight — an act of kindness that is still remembered with appreciation by the two evacuees. It was mid-January 1942, and it would be almost six weeks before Ruth Kydd, Rona Thomson and Stavranay Lamb would hear any information about the fate of their husbands who stayed behind to ensure that the QEA Engineering Station at Sourabaya remained operational for as long as possible.

QEA's flying boat *Corio* (VH-ABD) had been transferred to BOAC in exchange for their *Centaurus*, which had been impressed by the RAAF and was flying as A18-10. On 30 January 1942 *Corio* (carrying its new British registration G-AEUH) took off from Darwin under the command of Captain A.A. Koch on a special flight to bring back women and children refugees from Sourabaya. As they approached the island of Timor the aircraft was suddenly attacked and shot down by Japanese fighter aircraft, with the loss of ten passengers and three crew members. Captain Koch, First Officer Lyne and three passengers survived to give graphic accounts of the tragic event. Among the crew members killed was QEA's Senior Radio Officer A.S. 'Pat' Patterson, who (as stated previously) had flown as radio officer on the first delivery flight of the PBY5 Catalina from the United States to Australia.

By the beginning of February a 125,000-strong Japanese land force had amassed at the end of Johore Causeway, poised for assault on the island of Singapore. On 14 February an order was issued stating, in part,

'... as [Japanese] parachute troops have landed at Palembang, flying boats are not to proceed east of Calcutta or west of Batavia until further notice'. The following day Singapore surrendered to the Japanese, and the Empire flying boat route was severed.

Within days Sumatra also fell, and as the maelstrom of the Japanese advance moved southward, air raids on Batavia forced QEA to withdraw further; on 20 February the company's base of operation was transferred to Tjilatjap on the west coast of Java. Meanwhile, on the east coast of the island, Eric Kydd, Dave Thomson and Jim Lamb were living under trying wartime conditions in Sourabaya. On one horrific day Japanese aircraft unleashed a frenzied attack on the QEA base at Tandjoeng Perak. As there were no air raid shelters in the area, the three engineers had to lie in the gutter watching the enemy bombers overhead destroying everything in sight. When the raid ended they loaded all the flying boat spares they could salvage into the new 1941 Ford station wagon and, along with their few personal belongings and the Kydd's family dog 'Looky', set off overland to join Charlie Short, who by this time had arrived in Tjilatjap from Batavia. It was a considerable distance from the east to the west coast of Java, and the trip was not easy. For hygienic reasons they had always been advised not to eat from local food carts, but with no other form of sustenance to be found along the way, they were left with no alternative. They arrived in Tjilatjap to find that it had become a major evacuation centre, with flying boats operating a shuttle service to and from Broome in Western Australia. The Allied Supreme Command was evacuating Java, realising that it could not be held against the powerful Japanese forces, which had now begun to approach by sea from the east and were heading towards Sourabaya.

Lester Brain was sent from Sydney and put in charge of the operation at Broome, while Bill Bennett was made responsible for engineering operations and staff. Malcolm Millar of Mansfield & Co., who had looked after QEA's commercial interests in Singapore, oversaw the operation at Tjilatjap with the assistance of QEA's engineer Charlie Short. Three flying boat crossings were made into Broome before the regular shuttles began on 22 February. A total of ten journeys were made in all, on which eighty-eight refugees and 7 ½ tons of cargo were carried from Java.

On 28 February the last two shuttle aircraft, Corinthian and Circe, were ready to depart Tjilatjap, and Eric Kydd, Dave Thomson and Jim Lamb boarded Corinthian for the flight to Broome. Unfortunately, because of space restrictions on the aircraft, most of the spares they had brought with them from Sourabaya and poor Looky the dog had to be left behind on the wharf. Corinthian took off from the near-deserted harbour just before dawn, and Circe followed less than an hour later. That was the last anyone ever saw of Circe, its commander Bill Purton and crew. It disappeared without a trace, after sending a brief radio message about 200 miles out from Java, over the Indian Ocean.

With the exception of Charlie Short, all QEA engineering staff had now been evacuated from Java. Having seen the last QEA flying boat out of Tjilatjap, he proceeded with Malcolm Millar to Djogjakarta, and they were evacuated from there to Broome on a US Air Force Flying Fortress.

CHAPTER

7

DARWIN NIGHTMARE

Broome was considered less vulnerable to attack than Darwin and had therefore been chosen as the Australian base for the 'Tjilatjap Shuttles', despite the fact that conditions were primitive, there were no facilities for handling flying boats, and a 32 foot tide had to be contended with. In *Qantas at War*, Hudson Fysh observed: 'The flying boat mooring had to be laid in a deep-water channel more than a mile from the shore. The long jetty ran out for over half a mile, but when the tide was low, it stood thirty feet out of the mud and to get out to the moorings at such a time, it was necessary to climb down, walk through soft mud, and then finally wade out to a rowing boat carrying baggage and gear.'

Singapore had fallen, preparations were being made for the evacuation of Java, and as the Japanese began their unimpeded progression through the islands towards Australia, Darwin was placed on alert. For the first time there was a very real possibility that the enemy would strike directly at the Australian mainland, and Darwin was seen as the most likely point of attack. In a memo dated 13 February 1942 addressed to Norm Roberts (QEA's Station Engineer at Darwin), Works Manager Arthur Baird expressed his concern that Norm's wife, 'Honey' Roberts, had not been evacuated 'with the other ladies associated with our staff from Darwin'. He also wrote about the necessity of transferring most of the Darwin spares and materials to Broome, suggesting that Norm liaise directly with Bill Bennett at Broome and that he retain only the items he felt would be sufficient for a 'transit depot' at Darwin. 'There is a feature here,' he concluded, 'which, of course, I feel certain you have not overlooked — the fact that you might have to hurriedly remove from Darwin all or as many of our spares as possible, at some subsequent period.'

Despite the lesson of Pearl Harbor, Darwin, with its Air Force personnel, Army establishment and civilian population, was completely unprepared for the onslaught about to beset it. The Japanese had assembled a powerful task force at a point 220 miles northwest of Darwin and by 19 February, just four days after the fall of Singapore, was ready to begin its assault on Australia. The task force of seventeen ships consisted of four aircraft carriers, four heavy cruisers and nine destroyers. At the appointed time the aircraft carriers turned into the wind and launched an armada of 188 aircraft (fighters, bombers and dive bombers) for the less than one hour's flight, in complete radio silence, to Darwin.

Norm Roberts' detailed description of how QEA personnel survived the raids on Darwin makes compulsive reading, and in the interests of historical accuracy, is reproduced here verbatim:

> 'We had been driving with only narrow slits visible from our headlights for some weeks prior to the first bombing of Darwin and although the sirens had wailed the alert several times previously, there was no warning of the raids that first came on Thursday, February 19, 1942.
>
> There were twenty-four [QEA] people in Darwin at the time. This included two crews who were standing by the flying boat "Camilla", waiting further movement instructions before proceeding to Broome, where a rescue operation was developing for the evacuation of Dutch and Australian Nationals from the island of Java.
>
> On this fateful morning, Captains Crowther, Hussey and Gurney were at the Darwin Hotel, engaged in a new exercise of flight planning. As maps and other navigational equipment were required from the flying boats, I drove [First Officer] Don MacMaster down to the base to collect the necessary gear.
>
> "Camilla" was moored to the nearest buoy, which was about 200 yards from the wharf.
>
> During an earlier call to the base, I had arranged for the QEA launch "Naiad" to be waiting, and with five or six QEA staff aboard, we went out to the flying boat.
>
> We remained alongside for about ten minutes while the gear was gathered. We

P l a t e 1

▲ Flight Sergeant Arthur Baird overhauling a Beardmore Engine in Palestine in 1918. *(Australian War Memorial photo)*

▶ Qantas' first aircraft, the Avro 504K photographed at Mascot before its registration lettering (G-AUBG) was applied. *(Qantas photo)*

▼ The Qantas refuelling depot at Roma in Western Queensland. *(Qantas photo)*

Plate 2

▲ The arrival of the first Qantas Air Mail at Longreach on 2 November 1922. The aircraft is the Armstrong Whitworth FK-8. *(Linda Beard photo)*

▲ Qantas Bristol 'Fighter' (G-AUEB) rebuilt as a Bristol 'Tourer' by Qantas engineers at Longreach after it crashed into the railway goods yard in December 1923. *(George Roberts photo)*

◀ Qantas woodworkers Bob Struthers, Pat Andrews and George Dousha with the DH60 'Moth' (G-AUFU) in the Longreach hangar in 1927. *(Bob Struthers photo)*

▲ Arthur Baird (third from right) poses with the Longreach Aerodrome staff – all dressed in their 'Sunday best' – in front of the first Qantas-built DH50A 'Iris' (G-AUFA) on 18 August, 1926. *(George Roberts photo)*

▲ A DH50 wing under construction by Qantas engineers at Longreach in 1927. *(Bob Struthers photo)*

▲ The Qantas DH61 Giant Moth 'Apollo' (VH-UJB) at Longreach in 1929. *(Sandra Avery photo)*

▲ The Qantas engineering staff at Longreach pose for the camera aboard the Talbot utility in 1927. L to R: Pat Andrews, Dudley Wright, Bill Stone, Unknown, Jack Avery, Peter Turnbull, Unknown, Bob Ewings, Bill Flannigan, Harry Beadle, Jim Sanders, Bob Struthers, Hugh Adrian, Bob Jones, and Unknown. *(Sandra Avery photo)*

▲ Qantas engineering staff at Longreach with DH61 Giant Moth 'Apollo' (VH-UJB) in 1929. First and second Qantas apprentices Jack Avery and Bill Bennett are seen in the front row, far left and second from right respectively. *(Sandra Avery photo)*

▼ Qantas DH50A 'Iris' (now with registration VH-UFA) working the Flying Doctor Service in 1936, shortly before it was sold. *(Qantas photo)*

▲ Qantas aircraft in this photo are: DH60 'Moth' (VH-UGH), DH61 Giant Moth 'Diana' (VH-UJC) and DH50J 'Atalanta' (VH-UHE), outside the Qantas Hangar at Archerfield in 1932. *(Qantas photo)*

◄ Qantas DH83 'Fox Moth' (VH-URI) modified for use by the Flying Doctor Service in 1934. It replaced the original Flying Doctor aircraft (DH50 'Victory' –VH-EUR) which had been used by Qantas since 1928 to operate the service. *(Qantas photo)*

▲ Qantas DH86 'Canberra' (VH-USC) being examined by engineers in the Company Hangar at Archerfield in 1935. *(Qantas photo)*

▼ The inauguration of the first Air Mail between Australia and England at Archerfield on 10 December 1934. (L to R: H.R.H. The Duke of Gloucester, Captain Russell Tapp, Captain Bert Hussey, Engineer Dudley Wright and Engineer Norm Roberts.) *(Norm Roberts photo)*

▶ Flight Engineer Eric Kydd photographed at Koepang with a Qantas DH86 'Commonwealth Class' cabin bi-plane. *(Ruth Kydd photo)*

▼ Engineer Bill Gastrell testing a DH86 Gypsy Six engine on the Test Bed at Archerfield in 1936. *(George Roberts photo)*

◀ Qantas engineer Mark Webber servicing a Gypsy Six engine from a DH86 at Archerfield in 1935. *(George Roberts photo)*

▼ Works Foreman Dudley Wright with his aircraft 'The Wright-Mite' (which he built himself) outside the Qantas hangars at Archerfield in 1935. *(George Roberts photo)*

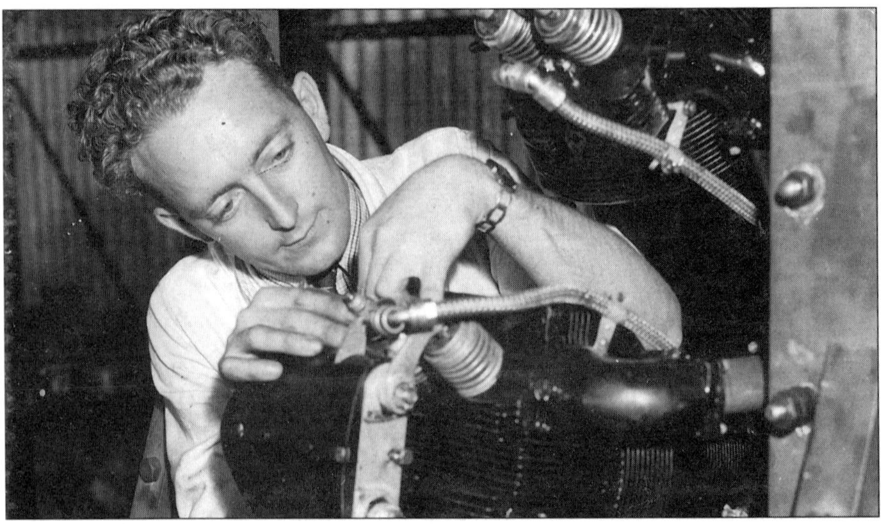

▲ Qantas Ground Engineer Ern Aldis working on one of the 920 Horsepower Bristol Pegasus 10C engines which powered the Short S23 'C' Class Flying Boats, shortly after his arrival in Brisbane from the United Kingdom in March 1938. *(Ern Aldis photo)*

▶ The crew of the first Qantas Flying Boat 'Coolangatta' (VH-ABB) upon its arrival in Brisbane on 2 April 1938. L to R: Bill Drury (Steward), Pat Paterson (Radio Officer), Bill Purton (First Officer), Captain G.U. 'Scotty' Allan, and Engineers Eric Kydd and Jan Aldous. *(Jan Aldous photo)*

▲ Imperial Airways' Empire 'C' Class Flying Boat 'Centaurus' (G-AUDT) moored on the Brisbane River and being examined by Qantas engineering staff in December 1937. Ruth Trickett (Secretary to Arthur Baird and later to become Mrs Eric Kydd) is standing on the wing. *(George Roberts photo)*

▼ The corrugated iron huts at Rose Bay which housed the original Qantas engineering workshop (right) and the catering section prior to the construction of the hangar complex. *(Jan Aldous photo)*

▲ Qantas Flying Boat 'Coorong' (VH-ABE) stranded on the rocks of the breakwater at Darwin in December 1938, where it was dismantled and returned to Short Bros in England. *(Norm Roberts photo)*

◀ The caption which accompanied this photo when it was published in a Sydney newspaper just prior to the inauguration of the Flying Boat Service from Australia under the Empire Airmail Scheme in August 1938 read... 'Six of the happiest faces in Sydney belong to Qantas engineers at the Rose Bay Air Base, who have been elected as overseers of the powerful engines that will drive the Empire Service Flying Boats.' (L to R: Engineers Col Sigley, Ern Aldis, Noel Miller, Norm Roberts, Bill Stone and Jan Aldous.) *(Ern Aldis photo)*

▲ Qantas Flying Boat 'Coogee' (VH-ABC) being towed up the new slipway at Rose Bay in 1939. *(George Roberts photo)*

▲ The new Qantas Engine Overhaul Workshop constructed in just 39 days at Kingsford Smith Airport at Mascot in 1939. *(Qantas photo)*

▶ The long-awaited Qantas hangar at Rose Bay, under construction in 1939. *(Jan Aldous photo)*

▼ An aerial view of the Qantas Flying Boat Base at Rose Bay showing the new hangar complex under construction in early 1939. *(Qantas photo)*

▲ The Qantas engineering staff at Rose Bay in 1941. Arthur Baird's working platforms, known as 'Rostrums' can be clealy seen in the background. *(George Roberts photo)*

▼ The vibration-free Instrument workshop in the Qantas Hangar complex at Rose Bay in 1939. *(Qantas photo)*

▲ Qantas engineers (L to R) Dave Thomson, Jim Lamb and Eric Kydd with their chauffeur and the Company Station Sedan in Sourabaya in 1941. *(Rona Thomson photo)*

▲ B.O.A.C's Flying Boat 'Centaurus' (G-AUDT), impressed by the RAAF for transport and sea patrol duties. It was repainted in heavy camoflage colours by Qantas engineers at Rose Bay and was given the new registration 'A18-10'. *(George Roberts photo)*

▲ Japanese bombs explode on impact across Kalang Aerodrome, Singapore, in February 1942. *(Commonwealth Department of Information photo 12455)*

▶ The Qantas hangar at Fanny Bay Aerodrome at Darwin, destroyed during the Japanese bombing raids on 19 February 1942. *(Qantas photo)*

▼ The Engine Overhaul Shop at Randwick in 1943, which was provided by the Department of Aircraft Production for the overhaul of aircraft engines for the Navy and the RAAF. Qantas supplied the expertise and the key staff. *(Ern Aldis photo)*

▲ Engine testing on the four wheel mobile test stands at Randwick in 1943. Note Ern Aldis' ingenious muffler system – the 44 gallon drum at the far left connected to the exhaust outlet on the engine. *(Ern Aldis photo)*

◀ Qantas engineers checking a US Army Air Force B-24D 'Liberator' Bomber engine installation at Archerfield in July 1943. *(Qantas photo)*

▼ Five US Army Air Force B-24D 'Liberator' bombers undergoing overhaul and modification in the open by Qantas engineers at Archerfield in July 1943. *(Qantas photo)*

P l a t e 1 3

▶ The Qantas Instrument Workshop in the Rose Bay Hangar Complex in the early days of World War II, prior to its transfer to a converted bus depot at nearby Double Bay in early 1942. *(George Roberts photo)*

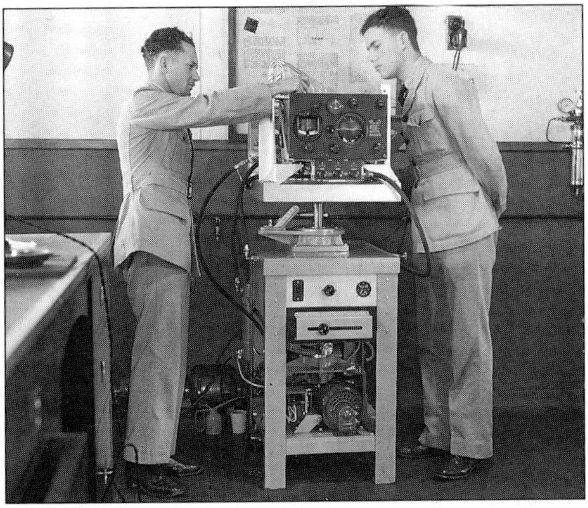

▶ The Sperry Gyroscope Company's 'Scorsby' at the Qantas Instrument Workshop at Double Bay. It was modified by George Roberts so that it could test simultaneously two Automatic Pilots and twelve Artificial Horizons or Directional Gyros in roll, pitch and yaw. *(George Roberts photo)*

▼ Qantas engineers overhaul a US Army Air Force B-24D 'Liberator' bomber in the half-completed Igloo Hangar at Archerfield in 1943. *(Qantas photo)*

▲ A Qantas Catalina in flight over the Indian Ocean in late 1943. *(Qantas photo)*

◀ (L to R) Station Engineer Eric Kydd, Engineer Dave Thomson and Coxswain Alf Back in Townsville during 1942. *(Rona Thomson photo)*

▼ The Qantas engineering staff at Townsville in 1942. *(Sammy Ames photo)*

▲ Qantas PBY-5 Catalina 'Rigel Star' running an engine test on the Swan River at Nedlands Catalina Base in 1944. PBY-5B Catalina 'Vega Star' (G-AGFL) undergoing overhaul in the Nose Hangar can be seen in the background. *(Norm Roberts photo)*

▶ An interior view of one of the Qantas Catalinas looking forward, showing some of the eight auxiliary fuel tanks. *(Norm Roberts photo)*

▲ Senior Station Engineer Norm Roberts and his assistant Col Sigley pose for the camera with the Qantas engineering staff at Nedlands Catalina Base in Western Australia outside the Nose Hangar and beside the Catalina PBY-5 'Antares Star' (G-AGIE) in early 1944. *(Norm Roberts photo)*

▲ Qantas PBY-5 Catalina 'Antares Star' (G-AGIE) alighting at Koggala Lake in Ceylon during 1944. *(Norm Roberts photo)*

◀ Qantas Foreman (Engines) at Nedlands, Alf Thompson, inspecting engine overhaul in the Nose hangar at Nedlands in 1944. *(Norm Roberts photo)*

▲ The first Qantas 'Liberator' LB30 (G-AGKT) photographed with the Western Operations Division's engineering and ground staff upon its arrival at Perth's Guildford Aerodrome on 3 June 1944. Senior Station Engineer Norm Roberts is standing beneath the wing at the far right. Note the first use of the 'Kangaroo Service' logo on the aircraft. *(Norm Roberts photo)*

▲ Qantas 'Liberator' LB30 (G-AGKU) came to grief when its nose wheel collapsed on landing after a test flight at Perth's Guildford Aerodrome on 16 October 1944. *(Norm Roberts photo)*

▶ The damaged nose section of the Qantas 'Liberator' LB30 (G-AGKU) undergoing repair in the hangar at Perth's Guildford Aerodrome during October 1944. The aircraft was returned to service on 13 November 1944. *(Norm Roberts photo)*

▶ The new Minister for Civil Aviation in the Menzies Liberal Government was T.W. White (later Sir Thomas), seen here inspecting the Qantas Instrument Overhaul Shop at Mascot in 1949 (left), with Qantas Instrument Foreman George Roberts (centre) and Ted Harper (right). *(Qantas photo)*

◀ John Upex and Arthur Littlejohn in December 1946. *(Betty Upex photo)*

▼ Engineers Jack Avery (left) and George Roberts (centre) with Qantas Publicity Manager E. Bennett-Bremner outside the main entrance to the Qantas Engineering and Overhaul Shop at Mascot in 1947. *(George Roberts photo)*

◀ Qantas Director of training Len Carolan (left-standing) supervises the Qantas engineers as they examine the landing gear system of the L749 'Constellation' at the Lockheed Training School in 1947. *(Frank Rowley photo)*

▲ (L to R) Lockheed Sales Administrator Walt Wayman, Qantas engineer Henry Williams, Captain G.U. 'Scotty' Allan, and engineers Doug Hudson and Norm Roberts pose by the fuselage of the first QEA L749 Constellation 'Ross Smith' (VH-EAA) in May 1947. *(Norm Roberts photo)*

▲ An aerial view of the Qantas hangar line at Kingsford Smith Airport, Mascot in 1947. *(Qantas photo)*

▶ The hinged faring on VH-EAV in the open position, showing the spare L749 'Constellation power egg' installed and ready for transportation. *(Ron Yates photo)*

▼ Qantas Avro 691 'Lancastrian' (VH-EAV), converted for use as an engine carrier, became known as 'Yates' Pregnant Pup'. *(Ron Yates photo)*

▲ Captain John Morton (left) and Ron Yates (Qantas's first professional Aeronautical Engineer) prepare to conduct a test flight in the Qantas 'Lancastrian' VH-EAV, which had been converted for use as an engine carrier. *(Ron Yates photo)*

◀ Qantas engineers change the engine of a Douglas DC4 'Skymaster' at Mascot in 1950. *(Qantas photo)*

▲ A Qantas DC3 undergoing servicing by Company engineers at Archerfield in 1948. *(Qantas photo)*

◀ Arthur Baird in 1949, just before his retirement from Qantas. *(Qantas photo)*

▲ Two DC3 Freighters undergoing overhaul and maintenance in the Qantas Hangar at Lae in 1957. *(Qantas photo)*

▶ The second Qantas DHA3 'Drover' (VH-EBR) above the Coral Sea on its delivery flight to New Guinea in April 1951. *(Qantas photo)*

▼ The interior of a Qantas DC3 freighter after modification for service in New Guinea at the QEA workshops at Archerfield. *(Qantas photo)*

▲ Qantas DHC2 'Beaver' (VH-EAV) in a scene from the Australian feature film 'Walk Into Paradise', which was made on location in New Guinea during 1956. *(Ron Yates photo)*

◀ The Qantas DHC3 'Otter' (VH-EAW) fitted with floats, operating the New Guinea internal routes in 1959. *(Qantas photo)*

▼ Chief Maintenance Engineer (Mascot Hangars) Ern Aldis (right) succeeded Doug Hudson as Chairman of the Apprenticeship Advisory Committee in November 1953. *(Ern Aldis photo)*

◀ Ron Yates arrived at Lockheed in February 1952 following his appointment as Qantas Resident Engineer. *(Ron Yates photo)*

▶ Of the nine original staff members who joined the company at Longreach who were still serving Qantas when this photo was taken in 1950, six were engineers. Back row (L to R): Dudley Wright (Engineer), Henry Williams (Engineer), Hudson Fysh (Chairman and Managing Director), Russell Tapp (Qantas Captain) and Phil Compston (Engineer). Front row (L to R): Arthur Baird (Engineer), Ida Isaacs (Secretary to Hudson Fysh), Bill Bennett (Engineer) and Jack Avery (Engineer). *(Sandra Avery photo)*

▶ Qantas Ground Engineers at the Lockheed Training School, Burbank during March 1954, prior to the delivery of the first Qantas L1049 'Super Constellation'. (L to R) Ken Barnes, Jeff Harris, Vaughan Sneesby, Roy Carr, Barry Watham, Charlie Eames, Harold Avery and George Foord. *(George Foord photo)*

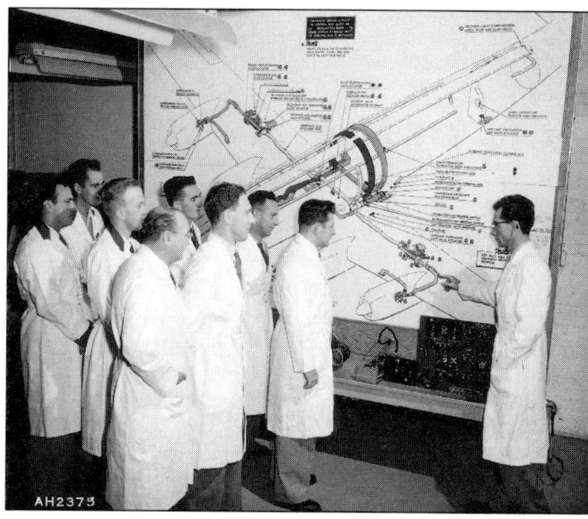

▼ The Qantas Engine Test Cell, built on the northern edge of the airfield at Mascot in 1956, to test both turbo-prop and piston type engines. It was the most advanced of its type in Australia. *(Qantas photo)*

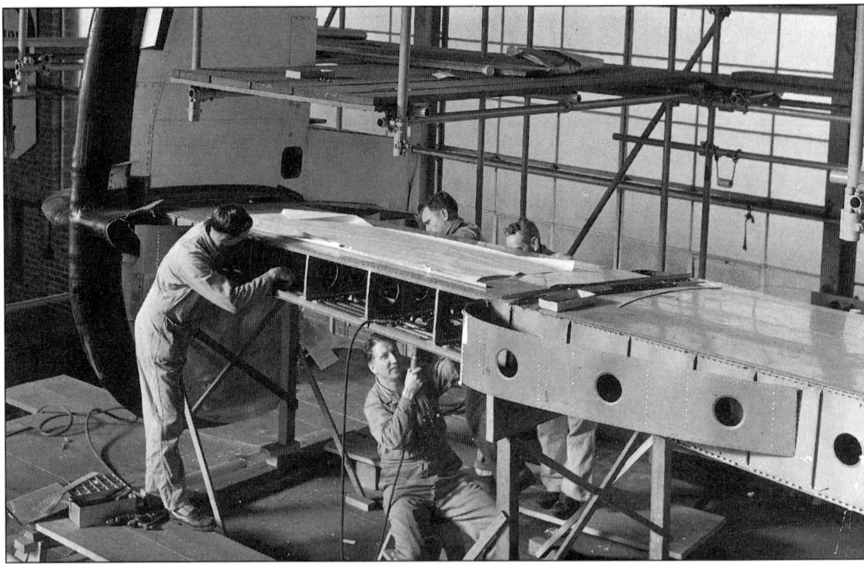

▲ The Qantas engineering staff at Mascot involved in the Tailplane Modification Programme which was carried out on the L1049 'Super Constellations' in mid-1956. *(Qantas photo)*

◀ A cylinder from an L1049 'Super Constellation' R3350 TC18 DA-1 Turbo Compound engine, overloaded in flight to the extent that it exploded out through the engine cowling. *(George Foord photo)*

▼ The Qantas L1049 Super Constellation 'Southern Dawn' (VH-EAD), modified to L1049 'G' standard by the fitting of wing tip fuel tanks by Company engineers at Mascot in 1957. *(Qantas photo)*

▲ The Qantas Prizewinning Apprentices for 1957 (L to R): Cec Morgan, Jimmy Chew Kim Seong, Lance Reece, Ken Gould (Best Apprentice Ground Engineer), Roger Critchley (Dux of the Qantas Apprentice School and winner of the W.A. Baird Shield), Jim Thurston, Eddy Grohe, Hans Haan and Neil Gaddes. *(Cec Morgan photo)*

▼ One of the options Qantas explored before placing its order with Boeing for seven 707 jet aircraft in September 1956, was the Douglas DC8. Pictured at the Douglas Aircraft Company's Santa Monica plant in July 1956 are (L to R) Qantas' Don Armstrong, Bob Walker (Resident Engineer, Burbank), G.U. 'Scotty' Allan (Assistant General Manager), Doug Hudson (Chief Engineer) and Ron Yates (707 Projects Engineer). *(Ron Yates photo)*

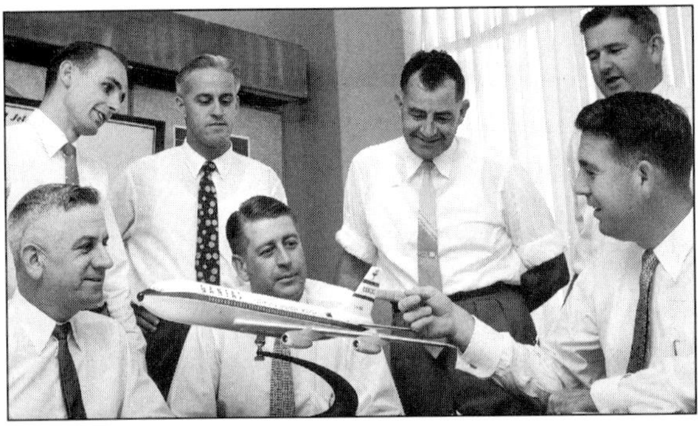

◀ The Qantas 707 Project Steering Committee in session at Mascot in April 1958. Standing (L to R): Max Hayworth, Captain Bert Yates, Harry Bewes and Bob Walker. Seated (L to R): Eric McFarland, Captain Ian Ralfe and Project Controller Ron Yates. *(Ron Yates photo)*

▼ The new Qantas Repair and Overhaul Section at Mascot in 1959. *(Bill Goodall photo)*

◀ The original piston engine Test Cell at Mascot was adapted to handle the General Motors Allison 501-D13 Turbo prop engines which powered L188C 'Electras' and C130 'Hercules' aircraft. Sub-foreman Jan Aldous (left) and Electrical Engineer John Cotton prepare an Allison engine for testing in 1961. *(Jan Aldous photo)*

▶ Qantas 707 V-Jet handling equipment on the tarmac outside the second Line Maintenance Hangar 191 in 1966. The Qantas DC3 (VH-EDC) and the Allison Engine Test Cell can be seen in the background. *(Sandra Avery photo)*

Plate 27

▲ The prototype of the mobile stair units built for Qantas by Hastings Deering at their Lidcombe plant in Sydney's west. *(George Roberts photo)*

▼ The former 117th Army General Hospital at Berrimah, 9 miles south of Darwin, was converted into a village to provide accommodation for both airline staff and passengers by Qantas in 1947. *(Qantas photo)*

▲ The passengers and crew of the Qantas Flying Boat 'Coriolanus' (VH-ABG) photographed on the wing prior to their departure for Singapore on 4 October 1945. Back Row (L to R): Engineers A.K. 'Kiwi' Stone, Norm Capel, Tony Koh, Senior Radio Officer G. Mumford, Mr W.M. Colvill, Traffic Superintendant G. Allen, Captains F.A. Reeve and K.G. (Keith) Caldwell, and Civil Aviation Airline Supervisor Mr. Christie. Front Row (L to R): Engineers Tony Lopes, Bill Sykes, Flight Engineer W.E. Wilcox, and engineer Roy Carr. *(Roy Carr photo)*

▶ Bill Bennett was promoted to the position of Superintendant Route Stations - Engineering in 1952. *(Qantas photo)*

▲ Qantas Station Engineers gathered for a conference in Sydney during May 1955. (L to R) Bill Barrie (Darwin), Con Gehrig (Development Superintendant), Bill Bennett (Superintendant Route Station Engineering), Eric Kydd (Chief Maintenance Engineer), Jim Lamb (Assistant to the Superintendant Route Station Engineering), George Dusting (Singapore), and Alan Keats (Manila). *(Qantas photo)*

▲ The Qantas 188C 'Electra' 'Pacific Electra' (VH-ECB) was placed into service on the Company's Far East Routes in December 1959. *(Qantas photo)*

◀ Qantas 707-138 'City of Perth' (VH-EBE) undergoing rudder maintenance in the open at Mascot from special scaffolding which was built on the airport apron in 1961. *(Qantas photo)*

▲ Apprentices from Singapore (L to R) Yeo Loy Seng, Lim Yew Wen and Gerald Goodenough undergoing training as ground engineers with Qantas in Sydney during 1960. Note the Qantas 40th Anniversary symbol on the tail of 707-138 'City of Sydney' (VH-EBB) in the background. *(Qantas photo)*

◀ The Pratt and Whitney JT3D-1 engine line at the Qantas Repair Overhaul Base at Mascot in 1962. *(Qantas photo)*

▲ Qantas 707-138s 'City of Brisbane' (VH-EBD) and 'City of Melbourne' (VH-EBA) on the tarmac in early 1961. The new Line Maintenance Hangar 131 is under construction in the background. *(Sandra Avery photo)*

▶ Senior Engineer Norm Roberts (Left) with Building Superintendant Ron McKinnell, standing in front of the new Qantas Line Maintenance Hangar 131 at Mascot in February 1962. *(Qantas photo)*

▲ The Qantas Hangar Line at Mascot in early 1961. The Sydenham-Botany goods railway line, which ran along the embankment behind the line of Super Constellations, had been removed to allow for the construction of the new 707 Line Maintenance Hangar 131. *(Sandra Avery photo)*

▲ The farewell luncheon for Sir Hudson Fysh with the 'Senior Citizens' of the Qantas Engineering Branch, held in the 707 mock-up at Mascot on 29 June 1966. Those seated represented 504 years of service to Qantas. (L to R): Bert Avis, Norm King, Bill Bennett, Jack Avery, Sir Hudson Fysh, Doug Hudson, John Gundry, Bob Banks, Ern Aldis, Harry Adby, Arthur Mitchell, Cliff Roberts, Sid Joseph, Phil Howson, Eric Kydd, Jack Macdonald, Ken Harris and Dick Ashover. *(Bill Goodall photo)*

◀ The Qantas Apprentice Training Centre at Rosebery in 1966. *(Qantas photo)*

▲ An interior view of the Qantas Apprentice School at Roseberry showing First Year Apprentices undergoing instruction in 1967. *(Jan Aldous photo)*

▲ The roll-out of the first Qantas 747-238B 'City of Canberra' (VH-EBA) at Boeing's Everett plant (Seattle) at midnight on 10 June 1971. Note the ten window configuration in the upper deck. Qantas' second 747-238B 'City of Melbourne' (VH-EBB) is the next aircraft visible along the production line. *(Ken Gould photo)*

▼ Qantas Chief Engineer Doug Hudson chairs one of his weekly engineering 'Clinic' meetings, which were held in the Conference Room in Hangar 96 at the Qantas Jet Base at Mascot in 1966. The meetings were held to keep regular tabs on the general 'health' of the Qantas aircraft fleet. Seated (L to R): Captain Phil Miller, Bill Hill, Ern Aldis, Bill Goodall, Doug Hudson, Ron Yates, Harold Hutton and John Mills. Portraits of Arthur Baird and Dudley Wright can be seen displayed on the wall at the rear. *(Ron Yates photo)*

▶ Qantas Chief Planning Engineer Arch Graham in May 1963. *(Arch Graham photo)*

P l a t e   3 4

▲ Six female First-Year Apprentices, chosen from forty applicants, receiving instruction at the Qantas Apprentice Training Centre at Rosebery in 1971. *(Qantas photo)*

▼ Using scale models, George Foord (Left) and Ray Golightly plan for the handling of the Qantas 747s at Kingsford-Smith Airport's International Terminal during 1970. *(Qantas photo)*

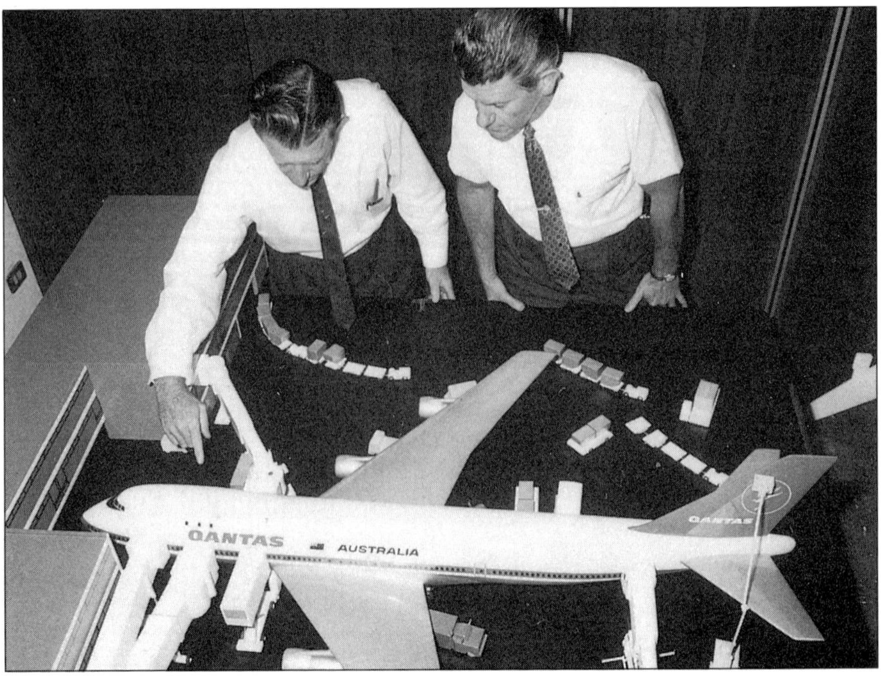

▲ The Tug produced for Qantas by the Fox Manufacturing Company in Sydney for the towing of both 747 and 707 aircraft. *(Qantas photo)*

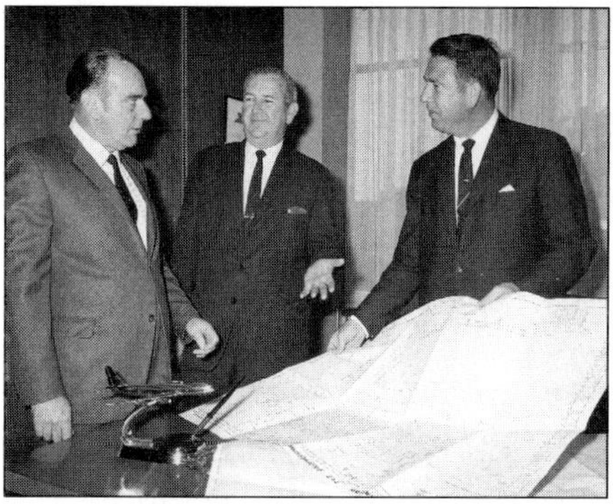

▶ Engineering Manager Doug Hudson discusses the Qantas 747 Engineering Programme with Technical Services Manager Bob Walker (Centre) and The Director of Technical Services Ron Yates (Right), prior to the arrival of the first Qantas 747 at Mascot in August 1971. *(Ron Yates photo)*

▲ Qantas Engineering Director Bob Walker (Left), Assistant Qantas Technical Representative – North America Michael Ryan (Centre), and Qantas Technical Representative – North America Ken Gould (Right) formally take delivery of the first Qantas 747-238B 'City of Canberra' (VH-EBA) on 10 June 1971. *(Ken Gould photo)*

▶ Qantas Chairman Sir Roland Wilson (Right) presents Jack Avery (Qantas' first apprentice) to Queen Elizabeth in the original Qantas Hangar at Longreach, on the occasion of the Company's 50th Anniversary in 1970. *(Sandra Avery photo)*

▲ The arrival at Sydney's Kingsford-Smith Airport of the first Qantas 747-238B 'City of Canberra' (VH-EBA) on 16 August 1971, escorted by the company's two HS-125 pilot trainers. *(Qantas photo)*

were under way again and heading back towards the wharf when suddenly the sea before us went up in a might roar.

Reacting to the impulse of self-preservation, we were quick to find the flights of bombers high overhead, in formations of seven, all approaching in line astern from the southeast. Moreover, we could plainly see that further sticks of seven bombs were well on the way down.

Deciding that our safety should best be gained by heading in the opposite direction, the "Naiad" was spurred into action and we made off for the open areas of the Bay.

During these very brief moments, however, the bombs continued to fall, and we were witnesses at very close hand to the death and destruction occurring at the wharf and to the ships nearby. It then became evident that bombs were also descending on the town.

At full throttle, we moved quickly for about half a mile, then the engine quit and the launch came to a stop. The trouble proved to be water in the petrol, but as the engine was hot, we worked with great difficult removing the carburettor. While we were working on the engine, the fast running-out tide carried us into the midst of the shipping and we could see that the "British Motorist", an oil tanker, had begun to settle and was heeling over. Other ships were also on fire.

The aerial activity had now intensified considerably with the arrival of a swarm of Zeros.

The carburettor back in place, we were under way again at full speed when one of the Zeros turned sharply and came at us from astern. We dropped speed immediately to foil his aim. Fortunately the Zero was low and his bullets went over our heads.

Realising that we ought to be less obtrusive, I took over the wheel of the "Naiad" myself and we got moving again at the lowest power we could use and yet make some headway against the tide. To minimise further attraction, we lifted the floor panels and, by having the remainder of the crew crouch in the bilges, I was able to position the panels on their backs, producing an impression that the boat was empty.

The ruse was successful and we ultimately reached the edge of the mangroves to the east of the Bay and greatly relieved, we came out of hiding. We learned very quickly that this move was a little premature as we came in for attention again from another Zero. His aim was high and the bullets sheared off some branches from the mangroves, which we used to camouflage the boat. With the "Naiad" now well concealed, but yet in about four feet of water, we all dispersed into the trees.

The disastrous result of the raid was now very grimly apparent and from our vantage point in the trees, the concentrated attack on the shipping and shore installations was in full view. By this time, the "British Motorist" was already on its side and oil was ablaze over a wide area of the water. Two ships at the wharf — the "Neptuna" and the "Barossa" — were also on fire, but the flying boat still appeared to be unharmed, although it was difficult to see in the smoke. It was probably due to the smoke that "Camilla" eventually survived.

At the same time, an American destroyer, the "Peary', was fighting back against a particularly heavy attack from bombers and fighters, but the end of the "Peary" came in one great explosion and the ship was engulfed in flame and smoke. We had a final glimpse of the stern going down, but when the smoke cleared, there wasn't anything of the ship in sight.

When the aerial activity subsided, we boarded the "Naiad" again and headed for the flying boat and were pleased to find it was completely intact. I later found two bullet holes in the tail, but machine gun fire had sunk the QEA passenger boat "Halcyon" at the next buoy.

The "Naiad" was now almost out of fuel, as it had been necessary to drain off some quantity of petrol to rid the tank of water. We beached the "Naiad" amongst the pearling luggers at the eastern side of the Naval oil tanks and made off to the [flying boat] base for more petrol. With enough fuel in the boat, it would then have been possible for us to tow "Camilla" to a safer anchorage away from the fires at the wharf.

Our arrival at the base coincided with an explosion of terrific force, accompanied by a great crescent of flame which seemed to envelop the whole area. A mass of debris and heavy objects lobbed all around us, but again we were very lucky. An artillery shell (very hot) landed about three feet from Don MacMaster and Roy Watson [one of the marine staff] as they dived under a railway truck for cover.

The explosion came from the "Neptuna" and was followed by the ship heeling over to settle on the bottom alongside the wharf. Afterwards, we could see that the stern section had been completely ejected into the mud on the opposite side of the wharf.

Until this event, the QEA building had escaped with only a few bullet holes, but then it just burst apart and the contents were strewn all around. The DCA building also suffered extensive damage and as I parked the Ford utility in front of the QEA building, the vehicle became almost lost to sight under wreckage from both buildings.

Somewhere also in the wreckage was the drum of petrol we needed for the "Naiad" and after some rummaging about, the drum, together with the necessary pumping gear was dragged out.

While extracting the Ford from the mess, we noticed that "Camilla" had gone and for a moment we feared the worst. Searching further afield, we were relieved to see the flying boat had somehow moved out ahead of the explosion and with engines running, was taxiing about at the edge of the Bay.

As we watched, "Camilla" took off. We later heard from the D.C.A. coxswain, John Waldie, that Captains Bill Crowther and Bert Hussey had boarded the aircraft from his boat. Following the departure of "Camilla", John Waldie continued to rescue survivors from the sea in the D.C.A. boat and did a memorable job.'

Note: Captains Crowther and Hussey had become aware that the *Neptuna* was loaded with explosives and had moved the aircraft out of range of the explosion before flying out of Darwin for Groote Eylandt, where they were able to refuel. They flew back to Darwin in radio silence, arriving at sunset to evacuate QEA personnel and equipment.

'Returning to the "Naiad" with the petrol, I left George Lake and Jack Wallace [from the Marine Staff] to gather up the equipment at the Base, while Roy Carr, Tom Baird and Roy Watson [also from the Marine Staff] made the launch shipshape and ready for work.

I had become anxious to know how my wife had fared in our QEA cottage out at Vesteys and since [First Officer] Don MacMaster wanted to rejoin the remainder of the crews, he and I got into the Ford and drove up to the town. There were very few people moving about, but the extent of the widespread damage caused by the bombs became evident.

There were no signs of the attack at Vesteys and I was happy to find my wife in the kitchen preparing tea for several soldiers, who had taken her into cover under a tree. Reassured that all was well and fortified by a cup of tea, I returned to the town and located other members of the QEA staff in the areas adjoining the Darwin Hotel.

I was looking for volunteers to accompany me back to the Base and Dawson Hare, the Acting Traffic Superintendent, was quick to accept.

To my dismay, I was to hear at the waterfront that the engine of "Naiad" had overheated and almost seized. To my discredit, I was also to learn that this had occurred only because we had refuelled the launch from a drum of white spirits and I had failed to notice the mistake. I knew we would be depending on "Naiad" for the time being at least, so we decided to remove the cylinder head to make a reliable repair.

Being aware that the Post Office had been wiped out, we realised that the telephones would no longer be working, so to follow the flight of "Camilla" it was necessary for us to drive out to the Fanny Bay Aerodrome where the D.C.A. "Airadio" Station was established. We wondered, too, if this was still there after the heavy attack we had seen in the area.

Leaving George Lake and Roy Carr working on "Naiad', I started out in the Ford for Fanny Bay with Dawson Hare. After calling briefly at the Darwin Hotel, we continued on and were passing the Oval when the second half of the raid began. Our warning came when "ack-ack" batteries went into action alongside us and we moved off quickly to find a gutter we could use if the bombs appeared to be coming our way. The formations of bombers [which had come from Ambon in the Celebes and were fifty-four in number] could be clearly seen flying at what I was later to learn was about 18,000 feet.

The town was spared on this occasion, but when the bombs hit the ground, we were sure that the RAAF Aerodrome had received a severe pasting. It was past midday when we got to Fanny Bay, where our attention was first attracted to the shattered state of the QEA Hangar. Fires were also burning in adjoining buildings, while the "rat-tat" of bullets exploding in an RAAF Store went on for some time.

It seemed fortunate then that the DCA building was still standing and had sustained little damage.

But talking with Arthur Tarlton of DCA, we found that it was not possible for us to get news of "Camilla" as the shock of a bomb blast had temporarily put the "Airadio" equipment out of operation and the supply of power had also failed.

An alternative source of communication at the Coastal Radio Station VID fortunately remained undamaged by the raid and although this installation was also mainly dependent upon the town supply of power, an emergency power plant kept it in operation. We were later informed through this system that "Camilla" was at Groote Eylandt and that Captains Crowther and Hussey were preparing to return to Darwin by dusk.

On examination, we found that the steel structure of the QEA Hangar was riddled with large holes from cannon fire and nearly all the sheeting had been torn off the roof and walls. As a specimen of this destruction, I removed a piece of the steel and shipped it aboard "Camilla" for viewing in Sydney.

By late afternoon, the "Naiad" was running smoothly again and because the services of every useful boat was required in the Harbour, I made "Naiad" available to the Navy.'

(Note: It is interesting to note that the Naiad was actually impressed by the Navy after the air raid, and in correspondence between Norm Roberts and Arthur Baird dated 26 February 1942 Norm reported the following: 'The "Naiad" was commandeered by the Navy immediately following the raid, for the worthy purpose of assisting survivors from the numerous wreckages of ships within the Harbour. Not content with this privilege, the Navy has since shown an utter disregard to our rightful claim of ownership and continue to use this launch as an emergency evacuation boat and as a tender to Naval shipping. I have obtained, with some difficulty, a concession to use the services of the "Naiad" to meet our own requirements when flying boats arrive but the launch remains under the control of the Navy coxswain. I have not requested the issue of a formal receipt for the "Naiad" as I felt that such action would end our chances of recovery. At the same time, the boat is receiving rather cruel treatment from the people concerned and, although they may be held responsible for any damage incurred, it is already showing signs of hard usage.

There appears little or no hope of holding civil property under the existing circumstances unless it is constantly guarded, so the "Naiad" may fare better in the end under controlled operations than if it were to remain at the mercy of individual Naval Officers who would undoubtedly use the boat to suit themselves.')

'Having confirmed that "Camilla" was on its way back from Groote Eylandt and was due in Darwin at dusk, I began to prepare for its arrival. The wharf had been breached by the bomb attack, severing the pipes which were used to load petrol into the Shell refuelling barge, making it impossible to onload fuel into the flying boat. Fortunately, Captains Crowther and Hussey had anticipated this situation and "Camilla" arrived with sufficient fuel for the return flight to Groote Eylandt.

Although we had become accustomed to working in blackout conditions, it was a tedious and time-consuming job to thoroughly examine the flying boat with only a pencil light coming from a torch. Therefore, it was close to midnight before we were satisfied that "Camilla" was ready.

I had arranged with the Navy for the "Naiad" to be returned during the night so that it could be used for the onloading of passengers and luggage in the morning, but when we arrived in full force at the Base, "Naiad" was not back. The only boat at the ramp was the dinghy we had used for our work during the night and we had to use it for loading until John Waldie came to our aid in the DCA boat.

Although a little later than originally intended, "Camilla" departed for Sydney without further difficulty.

To satisfy myself that all was well with the flight, I drove out to the "Airadio" Station at Fanny Bay. This was the only point of contact available to use and for months to follow, it kept us advised of aircraft movements.

Even so, on the evening of February 22, three days after the raid, a flying boat laden with evacuees from Singapore had come in from Broome without any warning and in need of fuel. The passengers, some of whom were BOAC staff, were taken up to the Darwin Hotel where they had to fend for themselves while the Shell Company's staff and ourselves had to overcome the problem of onloading fuel.

Fortunately, work had begun on the repair of the pipeline, but it yet needed a lengthy extension to reach the barge. By using a canvas hose, we took advantage of the high point of the tide and although it took all night to accomplish the task, we were finally successful in our effort. As the "Barossa" was still burning at the wharf, it was fortunate that the wind was in our favour.

When the aircraft departed that morning, my wife (the last woman to leave Darwin) was also on board. She had remained out of sight and safeguarded by friends since the raid.

Depleted in number after some of the staff were transferred to Broome, the remaining small band of engineers and Traffic Representatives moved into a house in the town and this became the QEA Headquarters in Darwin until the end of the War.'

As Japanese Zeros continued to make lightning raids on Darwin in the days that followed, it was thought that the enemy was planning to make a landing on the Australian mainland 'at the full moon'. Although this offensive never eventuated, most of QEA's stores and a number of engineering personnel were transferred from Darwin to Broome under instructions from Arthur Baird in Sydney. His memorandum to Norm Roberts dated 25 February 1942 was most specific: 'it will be Broome that will absorb the functions which were previously carried out at Darwin'.

Engineers Roy Carr and Tom Baird were sent from Darwin to render practical assistance at Karumba. A number of west-bound flying boats were being obliged to standover there because of the lack of mooring facilities at Broome and to comply with the new ruling that 'two or more flying boats should no longer coincide their arrivals at Darwin'.

In the late afternoon of 28 February engineers Eric Kydd, Dave Thomson and Jim Lamb arrived in Broome aboard *Corinthian* after their eight-hour flight across the Indian Ocean from Tjilatjap. As they walked along the jetty towards the township, a passerby called out 'G'day mates'. That simple greeting, more than anything else, assured them that they had at last arrived back on Australian soil. The following day, 1 March, Charlie Short also arrived, on the US Air Force B-17 that had flown him from Djogjakarta.

Bill Bennett, in charge of the engineering operation at Broome, had been joined by George Lake, Bob Timmins and Bob Jenkins, who had transferred from Darwin, as well as Wally Martz, Lloyd Hill and Arthur 'Speed' Robinson from Sydney. Arthur Robinson and Wally Martz, along with two BOAC engineers who had also arrived in Broome, accompanied the QEA shuttle aircraft to and from Tjilatjap as flight engineers.

On 2 March *Corinthian* left Broome, heading east under the command of Captain Steve Howard. The flight plan was to bypass Darwin and fly direct to Groote Eylandt, then proceed to Sydney via Karumba, Townsville, Gladstone and Brisbane. Travelling aboard as flight engineer was Eric Kydd. Dave Thomson and Jim Lamb remained in Broome to assist with servicing and refuelling the seventeen flying boats which were riding at their moorings in the harbour, anxious to depart as soon as possible the next morning.

Not long out of Broome, *Corinthian* developed engine trouble and was forced to make an unscheduled landing at Wyndham, where Eric Kydd worked through the night to make the engine serviceable. At 6:30 am on 3 March *Corinthian* flew out of Wyndham for Groote Eylandt, just two hours ahead of a force of Japanese Zero fighters which came in across the Timor Sea and struck the unsuspecting township, before heading southwest across the King Leopold Range towards Broome.

Two QEA flying boats were at Broome that morning: *Corinna* (G-AEUC) and *Centaurus* (A18-10, chartered to the RAAF). *Centaurus* was under the command of Captain Keith Caldwell, a former QEA pilot, and was being made ready for departure by RAAF personnel, while *Corinna* was being serviced and refuelled by QEA engineers. Engineer Bob Jenkins was on the wing attending to the refuelling of *Corinna* when, at 9:20 am, six of the nine Japanese Zeros suddenly appeared and began to attack. (The other three reportedly cruised at high altitude as a cover for the striking aircraft.) *Centaurus* was chosen as the first target; and as one of the Zeros swept in spraying explosive bullets through the top of the wings and into the fuel tanks, those RAAF servicemen who were able to, made a hurried escape in their rubber dinghy as the flying boat burst into flames and exploded. Then a second Zero began a similar attack on *Corinna*, shooting straight across the wingspan. With bullets already hitting and penetrating the wing all around him, Bob Jenkins took a head dive into the harbour. By the time he came back to the surface, *Corinna* was well ablaze and slowly sinking at her mooring. The QEA engineers inside the aircraft had scampered into a dinghy as soon as the attack had started, but finding that bullets were showering all around them they abandoned the dinghy and took to the water.

Dave Thomson and Jim Lamb were aboard a lugger with Bill Bennett heading out towards the aircraft when the Zeros swarmed in. They too were singled out for attention by an enemy aircraft, and so they took refuge in the workboat's dinghy, sheltering under the jetty as the lugger was sprayed with bullets.

The raid lasted only a matter of minutes, but it inflicted enormous damage to the defenceless base. A total of sixteen flying boats were sunk at their moorings, most of them Dutch Catalinas. Only one aircraft, a Dutch Dornier flying boat, had been able to fly out of Broome before the raid started. Having achieved their objective in decimating the flying boats, the Zeros then turned their attention on the Broome aerodrome, where a further six military aircraft, including two US Air Force Flying Fortresses and two Liberator bombers, were destroyed on the ground. One Liberator had succeeded in taking off just as the raid began but was attacked and fell a blazing wreck into the sea about 8 miles off the coast.

Bob Jenkins found a small dinghy that had come adrift from one of the luggers and climbing aboard, he set about gathering survivors from the sea, which was now ablaze with burning fuel. Two luggers that had been used for refuelling were surprisingly found undamaged and now became engaged in rescuing survivors, some of whom were badly wounded and dying in the water. They were mainly Dutch women and children, but there were also a number of RAAF personnel, who had survived the destruction of A18-10, aboard the lugger.

When the raid ended, Jim Lamb and Dave Thomson climbed out of their dinghy and up onto the jetty, where they were able to see for the first time the full extent of the devastation. As one of the luggers carrying survivors pulled in to the jetty, the two men quickly jumped aboard and tried to revive several people who had apparently drowned. Those who were still alive were transported in the little train that ran along the jetty to the Broome Hospital. Although there were no casualties among the QEA staff, there were many narrow escapes. Engineer Wally Martz was already in the hospital when the raid began; he had fallen victim to dengue fever, as had a number of people in Broome, and had been diagnosed as being too sick to work.

Shortly after the attack, the QEA flying boat *Camilla* arrived at Broome from Wyndham, under the command of Captain Eric Sims. The aircraft was carrying a group of passengers from the bombed MV *Koolama* and was actually flying an inland detour between the two towns while both were being raided. Alighting at Broome and finding burning wrecks and no moorings, Captain Sims taxied *Camilla* up into the sheltered area of a creek, where Bill Bennett and Captain Lew Ambrose hand-pumped enough fuel aboard to allow the aircraft to fly further south to Port Hedland.

Most of the QEA engineers remained in Broome for another four days, servicing the steady stream of flying boats and land planes which were passing through with evacuees who were being flown south to Perth via Port Hedland. Some attempt was made to carry out salvage operations on the wrecked flying boats in the harbour, but most of them were too badly damaged and little could be saved.

At about 9:00 am on 4 March a survivor from the Liberator bomber that had been shot down in flames over the sea the previous morning, stumbled into the hangar at Broome aerodrome. He had been in the water for more than thirty hours, finally reaching the shore soon after dark. A search and rescue operation was mounted immediately, and before dawn the next morning Captain Lester Brain set out from the Broome jetty in a refuelling launch, with Bill Bennett, an RAAF doctor, Flight Lieutenant Smith, and Carseldene, the Shell Company's coxswain. The operation, described vividly in E. Bennett Bremner's book *Front Line Airline*, was as exciting as it was dangerous for those aboard the launch.

> 'In the dim light, a strong tidal rip was running against a heavy wind. The launch being low in the water and not made for open sea work, water splashed in freely over the bow. It became necessary to start baling. Conditions got worse and a nasty swell began to make up. Waves broke freely inboard and the launch became almost uncontrollable.
>
> All hands baled furiously. A big green comber broke over the bows and completely swamped the forward part of the boat, stopping the engine and lifting the stern out of the water. The boat commenced to flounder.
>
> Desperately, in water up to their waists, Captain Brain and two of the others lifted some spare drums of fuel each weighing about 400 pounds, and heaved them overboard. After they had got rid of three, the boat righted itself and by vigorous baling was again got under control.
>
> By daybreak the engine was going again and the launch battled its way to the locality where the "Liberator" had been shot down. Various items of wreckage, lifejackets, burned cushions, pieces of uniforms and clothing were found during the morning, but there were no signs of any survivors.
>
> The launch covered an area of about 20 miles and the beaches were also searched. It was a very weary boat's crew that returned to Broome at noon that day.'

There was nothing further for the QEA engineers to do in Broome, so the next two days were spent gathering valuable spares and stores and making them ready for transport to Sydney. In the late afternoon of 7 March *Camilla* left Broome with all the remaining QEA staff and spares on board, flying south to Perth via Port Hedland.

After arriving in Perth the next day *Camilla* made one final flight back to Port Hedland under the command of Captain Frank Thomas, with an emergency cargo of dynamite needed for demolition purposes. Bob Jenkins also made the trip as flight engineer. Captain Thomas' report of the flight reads (in part) as follows:

> '... No engineering difficulties — encountered rather heavy bilgings. Engineer Jenkins did an excellent job of work.'

Hudson Fysh takes up the story in *Qantas at War*:

> 'In Perth after the last trip of *Camilla* up the coast, [Bill] Bennett was confronted with the need to coax a few more hours of what was now a sick boat for want of proper servicing. Bilging was frequent and Bennett pumped two hundred gallons of sea water out of her one night. There was metal in most engine filters and there were also a few feet off the end of the starboard wing'. [Note: This had occurred when *Camilla* was being taxied up the creek at Broome shortly after its arrival following the raid; the starboard wing had hit the mast of one of the luggers, and the extreme end of the wing tip had been sheared off.] 'We were still operating under civil orders and Bennett refused to sign *Camilla* out as airworthy for any further work up the coast.'

It is reported that both Captain Thomas and Bill Bennett paid high tribute to all the help they received from harbourmasters, oil companies and refuelling crews throughout *Camilla*'s operations. They particularly mentioned the Sea Scouts at Pelican Point, Perth, 'without whose boats and launches, ship-to-shore transport would have been virtually non-existent'.

Captains Eric Sims and Frank Thomas flew *Camilla* from Perth to Albany, with Bill Bennett and Arthur 'Speed' Robinson travelling as flight engineers. Then, 'after a lot of humouring and some anxious moments', they flew across the Great Australian Bight to Sydney via Adelaide and Geelong, reaching their destination on 18 March 1942.

Dave Thomson and Jim Lamb, along with the rest of the QEA contingent, returned to Sydney by a slow, roundabout route aboard a troop train. They arrived at Sydney's Central Station and were reunited, after just on two months' separation, with their anxious wives before commencing a well-deserved two-month leave.

Within days of the 'blitz on Broome', Japanese reconnaissance aircraft began flying down the east coast of Australia and were seen over Townsville. When news of this reached Sydney, Station Engineer Ern

Aldis was instructed to relocate the entire QEA Flying Boat Base from Townsville to Bowen, approximately 90 miles to the south. But by the end of April 1942 the situation in the north appeared to have cooled down, and so the operation was transferred back to Townsville. (It wasn't until December 1942 that Japanese bombs actually fell on Townsville.) In May 1942 Eric Kydd was appointed Station Engineer at Townsville and Ern Aldis returned to Sydney for reassignment.

Because of the rapid Japanese advance towards Australia and with memories of Darwin and Broome clearly in mind, QEA considered the possibility that its flying boat base at Rose Bay could also fall victim to enemy attack. As a result, the company set about finding a suitable site for a reserve flying boat base. Lake Boga, near Mildura in Victoria, was judged to be ideal in every way and, being fed from the waters of the Murray River, it was kept at a constant level. Although the QEA launch *Naiad* was earmarked for use at Lake Boga, it was still under Navy control in Darwin, with little chance of being released in the near future; another launch was therefore bought. An existing jetty on the lake was strengthened, spare engines and stores were sent from Sydney, and refuelling facilities and fuel supplies were installed by the Shell Company. Plans for a slipway and hangar were drawn up, and Bill Bennett was appointed supervisor of the whole operation. As events were to prove, QEA never had to use Lake Boga, but Bill Bennett is reported to have regarded it as a great rest-cure after his hectic few months in Singapore, Batavia and Broome. It eventually became a large and important RAAF flying boat base and overhaul centre, and for a considerable time it operated under the command of Scotty Allan.

Interestingly, an air raid shelter (which Arthur Baird described in correspondence as an 'eminently suitable structure') was built by the engineering staff at the southern side of the Rose Bay Flying Boat Base, just off the edge of the concrete apron. Excavated through layers of sand and shell to a depth of about 20 feet, it was large enough to accommodate all the QEA staff at the base. The trench was covered with timber, which had been lined with Malthoid, and because sand bags were difficult to acquire, the entire shelter was overlaid with 10 feet of sand that had been dug from the trench. Fortunately, the air raid shelter never had to be used by company personnel — not for the purpose it was intended anyway!

CHAPTER

8

AN ESSENTIAL INDUSTRY

With the fall of Singapore and breaking of the air link between Australia and England, QEA's airline operation was reduced to the Brisbane–Darwin inland route, which was being flown by the Lockheed 10A and the company's only remaining DH86, *Sydney* (the other four had been impressed for use by the RAAF). Qantas Empire Airways was now an international airline without an overseas route, and the agreement with BOAC was considered suspended from 19 February 1942, the day Singapore surrendered to the Japanese.

By May 1942 the company was left with only two Empire flying boats, *Camilla* (VH-ADU) and *Coriolanus* (VH-ABG). Three had been lost as a result of enemy action even though they had been flying as civilian aircraft (*Corio* G-AEUH, *Circe* G-AETZ and *Corinna* G-AEUC), and two more had been destroyed while under charter to the RAAF (*Centaurus* A18-10 in the attack on Broome and *Coogee* A18-12 in a landing accident at Townsville). *Corinthian* (G-AEUF) hit a submerged object during a night landing at Darwin on 22 March, and it too was completely wrecked.

Camilla and *Coriolanus* were now being operated for the Department of Civil Aviation (DCA), carrying military loads mainly between Sydney, Brisbane and Darwin. The DCA had assumed financial responsibility for the QEA operations, which, together with the planned expansion of engine and aircraft overhauls and repairs for the Department of Aircraft Production, was to be on the basis of 'cost-plus'. As the new situation developed, a great deal of work was carried out and paid for by the United States forces through the DCA.

During May 1942 QEA became involved in a spectacular air rescue operation behind Japanese-controlled lines in the highlands of New Guinea. Some ninety Australians, residents of New Guinea, had been cut off at Mount Hagen and were expected to be taken prisoner at any time. Two of QEA's DH86 aircraft, *Melbourne* (VH-USF) and *Canberra* (VH-USC), were loaned back to the company by the RAAF to carry out the rescue mission. They were hurriedly fitted with long-range fuel tanks and altitude controls at Archerfield and flown to Horne Island (on the tip of Cape York Peninsula), which was to be used as the base of operations. Between 18 and 24 May a total of eighteen flights between Horne Island and Mount Hagen were undertaken by QEA pilots Denny, Simms and Nicholl, and Guinea Airways' pilot Tommy O'Dea. Douglas Chambers and Robert Carswell, from QEA's engineering staff at Brisbane, accompanied the flights as flight engineers. Of the ninety people reported waiting to be evacuated from Mount Hagen, seventy-eight were brought out safely to Horne Island without loss or damage to crew or aircraft. (History has not recorded what happened to the other twelve.)

Dudley Wright was appointed Assistant Works Manager in May 1942, and this confirmed his position as QEA's second most senior engineer. While the future of the company's flying operation remained uncertain, the Rose Bay flying boat base was being swamped with work for the RAAF and the US government. In a memorandum to QEA Chairman Fergus McMaster, Works Manager Arthur Baird wrote:

> 'The requirements and pressure on this Flying Boat Base increases daily. This is brought about by the pressures of our own two flying boats, the RAAF flying boats, RAAF combat marine aircraft, the USA marine fighting aircraft and USA combat aircraft, together with the ever-increasing RAAF Rathmines [flying boat] operations and the Dutch marine aircraft requirements.'

Further compounding the pressure was an acute shortage of licenced ground engineers, a consequence of aircraft repair and maintenance work being undertaken for the Department of Aircraft Production (DAP). The

DAP had provided facilities at Randwick in Sydney and Moorooka in Brisbane for overhauling Navy and RAAF aircraft engines, and QEA had supplied the expertise and key staff. Jack Avery headed the Moorooka operation (at the outbreak of war he had been called up by the RAAF but was discharged a week later to continue essential work with QEA). Ern Aldis was Works Superintendent at Randwick. The facilities provided by the DAP included the buildings, all necessary tools and equipment, and the other staff.

At Randwick there were about 250 male and 60 female workers. They had been given six weeks' intensive training, after which they were placed in the workforce with a piece of paper which stated that they were successful 'dilutee' tradesmen. As Ern Aldis recalls, 'We had housewives, bricklayers, tennis coaches, people from all walks of life. What they lacked in depth of experience, they made up for in application to the task and they were certainly good enough to perform the jobs they were assigned.' The key personnel came from QEA's Engine Overhaul Centre at Mascot, which at that time was under the control of Arthur Littlejohns. John Upex, who proved to be a very capable engineer, was selected to be Works Foreman at Randwick.

They worked fifty-six hours a week, overhauling nine engine types, mainly Warner Scarab's, Armstrong Siddeley Cheetah IX's and Genet-Major's, Bristol Pegasus' and the occasional Pratt and Whitney. Testing of engines was carried out in the open on four-wheel mobile test stands. Randwick had three test stands, which were operated by Joan Kemp and Ted Raynor. The constant noise from engines under test, sometimes three engines simultaneously, annoyed the neighbours greatly and when they began to complain, Ern Aldis set about designing a muffler system to reduce the noise level. 'I took a forty-four gallon drum and with an eight inch flexible tube, I connected it to the exhaust outlet on the engine. I then made a smaller, inner container, placed it inside the drum and packed the area between it and the inside wall of the drum with "rockwool". With perforations in the metal surfaces, the noise was cut down enormously and we were able to continue through to the end of our operation without further complaint.'

Hudson Fysh described the Randwick operation as 'Our star undertaking' and in his book Qantas at War stated, 'Randwick engine overhaul shop in its day was the most impressive overhaul shop I have ever looked over, except, perhaps the United Airlines shop in San Francisco which was fabulous in its modern equipment...' The Randwick facility is reported to have overhauled in excess of 700 engines during its period of operation for the Department of Aircraft Production, and at its peak was processing sixty engines per month.

Female workers at Randwick performed a variety of onerous tasks in the workshop. Ranging in age from early-twenties to mid-fifties, most of them were housewives, and all of them worked very hard — until about four o'clock in the afternoon. Ern Aldis noticed that a number of his female staff would disappear into the ladies' locker room in the late afternoon each day, and he decided that he should find out what was going on. Arranging for the nursing sister to clear the way for him, he accompanied her into the locker room and was confronted by the sight of his female workers shelling peas and peeling potatoes for the evening meal. War or no war, their husbands and children at home had to be fed. Delivering what he considered to be a stern reprimand for carrying out household chores in company time, he nonetheless decided to turn a blind eye in future and allowed the practice to continue.

General Douglas MacArthur had arrived in Brisbane and set up headquarters at Lennons Hotel. He was soon followed by the American 5th Air Force, and QEA was placed under contract to overhaul the Wright Cyclone engines on their Boeing B-17 Flying Fortress aircraft. By 18 March 1942, the total B17 strength in the western Pacific region was twenty-six, all of which were based in Australia.

Lester Brain was placed in charge of QEA's operations in Brisbane, and Henry Williams was made Works Foreman at the company's hangar at Archerfield. To allow for the ever-increasing workload, it had been enlarged in late 1941 (under the supervision of the Aircraft Production Commission) and became known as Hangar No. 5. QEA had set up the engine annex of the Archerfield hangar to overhaul Armstrong Siddeley Cheetah IX and X engines and Gipsy Majors for the RAAF, but with the added influx of the Wright Cyclone engines from the US Air Force, the annex became crammed and engines began to flow out onto the hangar floor. It soon became apparent that the enlarged hangar complex was not going to cope, and so the Department of Aircraft Production set up another engine overhaul centre at Moorooka, to be operated by QEA on the same basis as the facility they had established in Sydney.

The DAP's Moorooka workshop was, like Randwick, in a residential area on the outskirts of suburban Brisbane about 4 miles north of Archerfield aerodrome. However, unlike Randwick, the testing of engines was not carried out on site at Moorooka; they were trucked to QEA's No. 5 Hangar at Archerfield, where they were run on test benches by Kit Brown and Connie Jordon (the first woman in Australia to hold a ground engineers licence, and the first female engineer to be employed by an Australian airline.

Before his appointment as Works Superintendent at Moorooka, Jack Avery had been sent on an extensive inspection tour of the aircraft factories and workshops in Great Britain and the United States to study tooling and equipment requirements for the large-scale production of modern aircraft engines. In England he found that the factories were operating with antiquated equipment and were a maze of overhead shafting and belts running down to all the machines. He marvelled that aircraft such as the Lancaster bomber could be produced in England under such circumstances. By comparison, the aircraft factories in America were of the latest design; the floors were clean, and each workbench was equipped with individual tooling and self-contained electric motors.

Despite his age (he was only twenty-one at the time), Jack Drinkall was appointed Works Foreman at Moorooka. This appointment was made with some reservation, but because most of the older, more experienced engineers had been taken to Rose Bay when QEA transferred its headquarters to Sydney in 1938, the company was left with little alternative. Jack tried to grow a moustache to make himself look older, but he was light-haired and the ruse failed to make any impression.

Arthur Baird sent for Jack Drinkall to come to Sydney: 'I want you to go out to Commonwealth Aircraft Corporation at Lidcombe, son, then to Technico, De Havilland's at Alexandria and James Kirby's at Camperdown. You will stay at each for as long as it takes to learn what you can about quality control at these places, then go back home and set up a quality control system to suit our line of business. We will have a retired engineer (Bob Jones) starting at Moorooka as your Chief Inspector, and for liaison purposes you will have a Wright Factory representative and a US Air Force sergeant to chase up parts, etc.' (The Commonwealth Aircraft Corporation was building Pratt and Whitney engines under licence at Lidcombe, while Technico were making generators, starters and coils. De Havilland's propeller shop was at Alexandria, and at James N. Kirby crankshafts were being manufactured for Gipsy Major engines. It was at James N. Kirby that Jack Drinkall first observed the dynamic balancing of parts; QEA had always used static balancing procedures, where propellers were balanced on a knife edge and given an extra coat of varnish on one side to get the balance right.)

Jack Avery spent a great deal of his time designing special tooling and having it made and set up for the Moorooka workshop, leaving Jack Drinkall responsible for the progress of engine overhaul and staff training. As was the case with Randwick, the staff comprised a number of key QEA personnel (including Ken Barnes, George Riek, Graham Marriott, Mick Wass and George Foord) and 'dilutee' tradesmen and women who had been drafted 'off the street' and shown how to use a file and a micrometer at a school set up by the DAP. They had no expertise on aircraft engines and had to be trained on the job.

The overhaul manual for the Wright Cyclone engines was a restricted US Air Force publication, and was not available for use by 'civil' operations. However, it was obvious that QEA would not be able to operate without one. By means that are still unclear to this day, Jack Drinkall acquired a single copy of the coveted manual and tore it into sections, distributing each part to the relevant area of the engine overhaul line.

Sandblasting was being used to clean engine parts after they had been de-greased, but it was found that some parts were being damaged by the force of the sand. Seeking a less-abrasive alternative, Jack Avery experimented with bird seed but found that it caused the parts to become sticky. Next he tried kibbled wheat, and it worked so well that it was used constantly as a cleaning agent right through to the end of the war.

On one occasion, the Americans brought three Japanese Kasei twin-row engines to Moorooka and asked that they be cannibalised and rebuilt as one engine, to enable them to get performance figures. ('Kasei', Japanese for the planet Mars, were engines built by Mitsubishi). The engines had been salvaged from an island that had been taken by the Americans, and Jack Drinkall gave a charge hand and his team in the dismantling section the job of stripping and cleaning them. But the men refused to touch the engines, and upon investigation Jack found that the entire inside of the engines had been flyblown and were a crawling mass of maggots the size of silkworms; apparently the Japanese did not use mineral oil in their engines but a substance based on vegetable oil or animal fat. The men accepted an offer of two shillings per hour 'dirt money' to work on the engines after they had been flushed with phenyl, and a single engine was eventually built to suit the Americans' requirements.

Moorooka had a Wright Cyclone representative named Louis Stankovich allocated to work on site, and he was able to make direct contact with his factory in the United States when a problem arose that he could not solve himself. There were also two US 5th Air Force personnel permanently stationed at Moorooka, and their function was to liaise between QEA and the US Air Force. The workshop processed more than forty Wright Cyclone engines per month, as well as the odd Gipsy Major and Pratt and Whitney engine.

At one stage they were also overhauling engines for the Netherlands East Indies Air Force, and it was during this period that Henry Williams, QEA's Works Foreman at Archerfield, came to Moorooka one Sunday with a high-ranking officer of the Dutch Air Force. They wanted to check on the progress of the engines. The caretaker at Moorooka was a somewhat eccentric Englishman who had been a swagman, and Jack Avery (for reasons best known to himself) had given him the job of Watchman/Caretaker. He lived on site in the front Security Office. The factory building was surrounded by a very high fence, and the only entrance was through two large gates at the front. When Henry Williams and the Dutch Air Force Officer asked to be admitted, the caretaker refused to open the gate.

'Do you know who I am?' asked the exasperated Works Foreman. 'I'm Henry Williams.'

'I don't care if you're the Prince of Wales, Mr Williams,' replied the caretaker, 'No-one comes in here unless Mr Drinkall says so.' And they didn't get in.

Henry Williams wanted him fired, but Jack Drinkall declined on the basis that he was probably the most efficient caretaker in the organisation — if he wouldn't let Henry Williams in, there was not much hope of anyone else getting in. So the English caretaker stayed on.

According to Hudson Fysh, 'the new engine shop [at Moorooka] under Jack Avery soon earned a first-class name for itself, the little label on each engine, "Overhauled by Qantas", being much sought after.'

By the end of January 1943 the company's operation had become totally non-commercial. The staff, which exceeded 900, was continuing to grow as approximately 80,000 manhours per month were being worked for the Department of Aircraft Production. The Rose Bay Flying Boat Base, under Arthur Baird and Dudley Wright, was the busiest QEA centre in Sydney, employing just under 400 people. George Williams and Bill Stone continued to supervise the DAP work at Rose Bay, while Jan Aldous was transferred to the company's new Igloo Hangar at Archerfield.

Available hangar space had once again become a problem at Archerfield. QEA engineering staff were working ten hours a day, six days a week, repairing and overhauling aircraft for the US Army Air Corps. Work on the B-17 Flying Fortresses had been supplemented in April 1943 by the arrival of a number of B-24D Liberator bombers requiring urgent modification to their forward armament. This work had to be carried out in the open, pending the completion of the Igloo Hangar, which was being built for QEA by the Department of Aircraft Production.

John Franetovich was a planning engineer with the Brisbane Maintenance Division of the DAP in 1943, responsible for the design of workshop layouts and installation of plant and machinery required to service aircraft. Describing the design and construction of the Igloo Hangar, he wrote:

> 'The trusses were made from pieces of timber nailed together and each truss section was bolted only at the heel and apex. The individual trusses were made up in two sections and then lifted onto heel blocks at each side. The apex points were raised and lowered until they met — their weight held them in position. Purlins were nailed to each adjoining truss and corrugated iron sheeting was attached to the purlins. This cladding had the effect of stiffening the whole configuration.

> The hangar had one weak design fault — if one end was open, the other end also had to be open, otherwise the structure would have exploded when the wind caused a build-up of pressure inside the igloo.'

So urgently needed were the Igloo Hangar facilities that QEA began using them before construction was completed, carrying out repair, overhaul and vital modifications to the B-24D Liberator bombers. It had been found that the early model B-24Ds were vulnerable to attack from front-on. Their forward armament consisted of a single half-inch calibre machine gun in the nose, which projected through a port, giving a much smaller arc of fire than the twin half-inch guns mounted in the tail turret. Supplies of tail turrets were sent to Archerfield, where QEA staff removed the original nose structure and modified the area to take the heavier weapons. At least forty twin half-inch machine gun turrets with bulletproof glass were fitted to the nose position of the B-24Ds during 1943.

On one occasion, US Air Force Brigadier-General Connell paid a visit to the Igloo Hangar and was being shown around by Jan Aldous. There were five Liberators being worked on at the time, and the Brigadier-General very quickly determined that there was an acute shortage of suitable tools and equipment required to do the job. The next morning a convoy of ten US Army trucks arrived at the Igloo Hangar with a load of equipment: two Ford V8-driven air compressors, a 240 volt power-generating plant, cases of rivet guns, electric drills, everything that was needed. Naturally, the rate of production increased out of sight, and a significant improvement in staff morale was also evident.

Not all of the 600-odd people employed at the Igloo Hangar were QEA employees; in fact, most of them worked for the DAP under Qantas direction. About seventy of them were women, six of whom were skilled sheet-metal workers, and two who were very good at engine maintenance.

As the war began to move closer towards Australia, QEA began assembling Lockheed P38 Lightning fighters (about thirty in all) and six Douglas A20 Boston light bombers at the Igloo Hangar, as well as repairing combat-damaged aircraft of various types that had come in from the battle areas in the north. One aircraft arrived with a top-secret Norden bombsight fitted, and in 'ten seconds flat' QEA's instrument engineer Brian Murton had it out of the aircraft and into the instrument shop, where it was examined in minute detail. He then reinstalled it in the aircraft. The following day an armed guard arrived and took the 'top-secret' bombsight away.

The need for still more combat aircraft prompted the setting up of a production line at Eagle Farm aerodrome in Brisbane, where QEA engineers assembled close to sixty Bell P39 Airacobra fighters for the US Air Force. (After the war General George C. Kenney, head of the US Air Force, wrote to QEA stating that the output of aircraft from the company's Brisbane operation had far exceeded their expectations, and he offered the opinion that it had had a marked effect on bringing an early end to the war in the Pacific.)

The QEA operation at Archerfield existed on two levels between 1943 and 1945: at the Igloo Hangar 'military' aircraft were repaired and overhauled under the supervision of QEA's Jan Aldous, whereas No. 5 Hangar, under Henry Williams, remained essentially a 'civil' operation. Engines and airframes from 'civil' aircraft required licenced ground engineers to sign them out, but with 'service' aircraft no such requirement was necessary since they did not fall under the jurisdiction of the Department of Civil Aviation.

QEA had been allocated a number of Lockheed 18 Lodestar (Military C56) aircraft, which were on loan from the US Army Air Corps and were being operated for the Directorate Air Transport under agreement with and direction of the Department of Civil Aviation. They were powered by two 1200 hp Wright Cyclone 1820-102A engines and carried Australian civil registrations (VH-CAA, VH-CAB, VH-CAF and VH-CAK) with the name 'Qantas' on the nose. They had begun intensive operations into New Guinea carrying supplies and evacuating wounded soldiers in November 1942. Because the Lodestars were being operated by QEA crews as 'civil' aircraft, their servicing and overhaul was being carried out by Henry Williams' team at the company's No. 5 Hangar at Archerfield. Connie Jordan was the only QEA ground engineer to hold a licence for the Lodestars, hence she was signing them out. As Jan Aldous recalled the situation,

> 'We thought, "if she's signing out the aircraft, she should be in charge". So we put her in charge and immediately all the men walked off the job — they weren't going to work for a woman.
>
> Henry Williams rang me at the Igloo Hangar and said, "you'd better bring the Union Representative over here, we've got a bit of a problem. They've all gone on strike." So I went across to the QEA Hangar with Dobson (the Union Rep.), filling him in on what was happening on the way over. When we got there, everyone was assembled outside the hangar and Dobson got up on a ladder and said, "I've never heard of such a dastardly act done by anybody — to put a team of able bodied men under the charge of a woman is unbelievable. I reckon it's probably the worst thing I've ever heard." He continued in this tenor for about five minutes and then said, "Where is the man with a licence on this aircraft and I'll see that he gets the job straight away." And there was dead silence — no-one else had a licence on the Lodestar. He just stood there and looked at them for a minute before issuing his final

> statement on the subject — "If not one of you men have enough brains to get a licence on this aircraft, you'd better go back to work for a woman then, and when one of you does get a licence, I'll make sure that man gets the job." And they all went back to work.'

Connie Jordon was a unique personality among the many outstanding individuals in QEA's engineering section. Not only was she the first woman in Australia to hold a 'C' and 'D' class Ground Engineer's licence (No. 1430), which would carry nineteen endorsements by 1953, but by the time she joined QEA at Archerfield in 1942 she had already gained her private 'A' pilot licence (No. 1800, in 1936) and had earned four degrees from Trinity College in London in pianoforte and violin. She also taught ballet and ballroom dancing. Then in 1944 she bought a 1937 model MG sports car and became an active member of the Queensland Motor Sporting Club, driving in car trials and road races. At that time she was the only woman in Queensland driving racing cars.

Shortly after her employment as a ground engineer at Archerfield, she was posted to QEA's line station at Cloncurry as assistant to Station Engineer Bill Williams. She spent two years at Cloncurry servicing the QEA Lockheed 10A *Inlander* as it flew the inland route between Brisbane and Darwin and also the new Flying Doctor aircraft, the DH84 Dragon (VH-URY, named *John Flynn*), which always had to be kept ready for an emergency flight at any time of the night or day.

Returning to Archerfield in 1944, she assumed responsibility for the testing and signing out of aircraft engines after overhaul. Her engine test 'laboratory' was a fibro shed about 10 feet square, which was near the bush at the far end of the aerodrome. Her test bench stood beneath the shed's only window and was often covered with slivers of glass from holes smashed in the window by small stones picked up by the propeller on the engine being tested outside; the engines sat on fixed mounts set in a concrete slab about 20 feet away from the shed. As her experience and expertise grew, her 'D' class ground engineer's licence became endorsed for both in-line and radial engines, including the Pratt and Whitney twin-row engines, and it is thought that she was the only ground engineer in Queensland with a licence endorsement for the whole Pratt and Whitney series engines and the entire Wright series.

No account of QEA's period as 'an essential industry' would be complete without mention of the contribution to the war effort made by the company's Instrument Overhaul facility at Double Bay. In early 1942 the QEA Instrument Workshop was transferred from the Rose Bay hangar complex to an old bus depot building on the corner of Bay and Cooper Streets at Double Bay. Initially the building housed both the Instrument and Engine Overhaul sections, but when the Department of Aircraft

Production provided facilities at Randwick for the overhaul of engines, the building at Double Bay was given over to the Instrument Overhaul section for its exclusive use.

George Roberts had chosen the Double Bay site in consultation with Arthur Baird when it was thought that the transfer was to be short term, pending relocation of the entire QEA harbourside operation in Sydney to Lake Boga. However, as the proposed undertaking at Lake Boga did not eventuate, the QEA Instrument Overhaul section remained at Double Bay until the end of the war. It wasn't long before the operation, under the supervision of George Roberts, began to grow. A mezzanine floor had to be built and a large cellar dug underneath the building to provide more space. Instruments were being overhauled for the RAAF, the RAF, the RNZAF, the United States, French and Dutch Air Forces, as well as QEA's own aircraft. Gyroscopic instruments, from the US Air Force in particular, were coming in in their thousands, and for a time this was more than the QEA staff could handle. Through the Department of Aircraft Production, the jewellers and watchmakers of Sydney were 'manpowered' to QEA and set to work overhauling aircraft instruments, while many women from nearby suburbs were employed and trained on a production-line basis to overhaul, calibrate and test all types of aircraft instruments for the duration.

On the mezzanine floor, a restricted area known as the 'luminising room' housed equipment for painting instrument dials with a radium-active paint called 22M Compound, which was supplied from the United States in lead-sheathed files. The luminising room's eight female operators, working four to a shift, each had to pass through an ultraviolet light in a specially constructed darkroom, where their clothing was inspected for even the most minute traces of luminous paint when they began and finished each shift.

All instrument overhaul equipment had to be built by QEA engineers at Double Bay, the most prominent of which was undoubtedly the Scorsby. Designed originally by the Sperry Gyroscope Company in New York to fully test an automatic pilot in roll, pitch and yaw, and a single item of blind flying equipment, the Scorsby was modified by George Roberts so that it would test two automatic pilots and either twelve artificial horizons or directional gyros simultaneously.

One of the constant problems to plague George Roberts was the supply of instrument spare parts. Most spares were stored at the RAAF Base 6SD at Dubbo in western New South Wales, but no-one at that establishment knew what the spares were for. The situation was brought to a head when six Lockheed Hudson aircraft had to lie idle at Sydney's Bankstown aerodrome, grounded and unable to fly because their exhaust gas analysis equipment was in the QEA Instrument Workshop at Double Bay and no spares could be obtained for them. To solve the problem, the Aircraft Inspection Directorate arranged for George Roberts to go to No. 6 Stores Depot at Dubbo under the pretext that instruments received and tested from Double Bay were unserviceable and required investigation, a pretext that provided entry for a civilian into a military complex. With the authority of Group Captain Rickard, the officer in charge of the base, George was assigned four WAFS and twelve air crew awaiting overseas postings, to move and open boxes containing spares for inspection and identification. George was able to locate all the spares needed to get the Hudsons back in the air, and they were issued directly to him on the spot. He returned to Sydney with them by train, carrying with him a 'bonus' batch of 28,000 bearings and some much-needed equipment that just happened to be lying around unused on the shelves. Output at Double Bay apparently increased ten-fold as a result of George Roberts' trip to Dubbo, especially after a similar exercise conducted at No. 7 Stores Depot at Drayton, Queensland, where a further 28,000 ball races, specially selected for gyroscopic instruments, were obtained.

As well as his primary responsibilities relating to instrument overhaul at Double Bay, 'Professor' George Roberts was also running a course for service personnel on maintenance and overhaul of Sperry blind-flying equipment and automatic pilots. (The nickname 'Professor' had been ascribed to George by E. Bennett Bremner, QEA's publicity officer and author of the book *Front Line Airline*.) The servicemen came to 'Professor' Roberts from the RAAF, the RAF (from Singapore mainly, before it fell), the RNZAF and the American, French and Dutch Air Forces, through the RAAF No. 5 maintenance group at Point Piper. 'The Long Sperry Course', as it was called, was a fulltime six-week course held in the QEA Instrument Workshop. All the service personnel who attended the school were instrument specialists, but none of them had any experience with gyroscopes. They were taught, six at a time, the rudiments of gyroscopic instrument overhaul, and by the end of the course most of them gained practical experience in the QEA workshop before returning to their bases.

QEA's total war production was a splendid record of achievement and a credit to all the engineering staff involved. In his book *Qantas at War*, Hudson Fysh revealed that '293 aircraft were repaired or overhauled and 2,814 aircraft engines were overhauled'. The figure shown for instruments overhauled was 24,328, but according to George Roberts' comprehensive personal records, the actual number of instruments subjected to thorough examination and repair between 1941 and 1945 was actually slightly in excess of 143,000. (While hostilities in the Philippines and at Gaudalcanal continued, US Air Force C54 transport aircraft were carrying all manner of aircraft instruments, including bombsights, into Mascot for transfer to Double Bay. No vouchers were used by the forces, and it became the responsibility of the QEA stores staff at Double Bay to raise the essential paperwork for endorsement by the US Air Force headquarters at 7 Wynyard Street, Sydney. This probably accounted for the large discrepancy in accredited recordings.)

On 21 December 1943 QEA's managing director, Hudson Fysh, sent a letter to (presumably) all senior engineering staff. The letter that engineer Jack Drinkall received, accompanied by a bonus cheque for £10 ($20), read as follows:

'CONFIDENTIAL

Mr J.C. Drinkall
Works Foreman
MOOROOKA

Dear Drinkall,

It is a pleasure to enclose a bonus cheque with this letter, which is a small mark of appreciation by the Board of Directors of this Company for all the hard work, devotion to duty and keenness you have displayed during the past financial year.

The bonus is something apart from salary and as you will realise its payment depends on the financial outcome of operations and other considerations.

Through all that the Company and the individual is going through during wartime and in the face of wartime conditions, wartime expansion and diffusion, we must keep united to ensure that the central core of QEA remains firm and also progressive to meet the future.

The Board and Executive thank you for your assistance and wish you a Merry Christmas and Peace through Victory in the New Year.

Yours sincerely,
(signed)
HUDSON FYSH
MANAGING DIRECTOR'

CHAPTER

9

SECRET OPERATIONS

In June 1943 QEA acquired a second Empire flying boat to supplement the wartime activities of its only remaining aircraft, *Coriolanus*, still operating under civil orders. When Singapore fell on 15 February 1942 and the Empire Flying Boat route was severed, one of BOAC's flying boats, *Clifton* (British registration G-AFPZ), became stranded in Sydney. Since an additional aircraft was needed to fly troops from Townsville into New Guinea, *Clifton* was impressed by the RAAF on 12 March 1942 and given a new identification A18-14. Then, on 26 June 1943, after *Clifton* was released by the RAAF to resume civil operations, it was bought by QEA on 1 July for £51,084 ($102,168) and allocated the Australian registration VH-ACD. Unlike all other flying boats that had formed the QEA fleet, which were Short S23s, *Clifton* was an S33 model, the type built to operate the Atlantic Ocean route. They were powered by the same four 920 hp Bristol Pegasus XC engines as the S23s but were constructed wider at the chine to allow for the weight of additional fuel carried.

Calypso had been lost in August 1942 when it crashed near Daru in New Guinea while flying patrol duties for the RAAF as A18-11. Of the four QEA flying boats impressed by the Australian Government, only one now remained, *Coolangatta* (A18-13), and the RAAF handed this back to QEA on 29 July 1943. By the beginning of August 1943 QEA's flying boat fleet consisted of *Coolangatta* (VH-ABB), *Coriolanus* (VH-ABG) and *Clifton* (VH-ACD).

When Eric Kydd became QEA Station Engineer at Townsville in May 1942, the north Queensland town was just beginning to assume a new importance as a strategic military location. The RAAF's 33 Squadron was stationed there, and was transporting troops to the northern war zone. But as military activity increased, QEA's *Camilla* and *Coriolanus* were pressed into service and, under the control of the War Transport Command, began carrying 'troops, complete units of special aerodrome Defence Corps, light anti-aircraft guns, ammunition bombs and equipment, as well as commandos and key Army personnel' from Townsville into the forward military areas of New Guinea. Return flights brought out the sick and wounded and occasionally a Japanese prisoner of war for interrogation at Army Intelligence Headquarters in Melbourne.

From August 1942 *Camilla* and *Coriolanus* maintained a daily service in each direction to Port Moresby and Milne Bay, on the eastern extremity of New Guinea, and fortnightly flights to Sydney for engine and airframe overhaul. In *Qantas at War*, Hudson Fysh reported that they kept '... to a regular timetable and [were] so much on the go that at one period, they reached the rate of just under four thousand hours per year, which was fantastically high for a flying boat. Behind this effort lay the work of the [QEA] Sydney engineering base and our Townsville engineers'.

The number of engineers required to operate from QEA's small corrugated-iron workshop at Townsville increased considerably when *Camilla* and *Coriolanus* began their daily charter flights into New Guinea. The QEA engineers known to have served in Townsville for varying periods between 1942 and 1945 were Eric Kydd* (Station Engineer), Albert 'Sammy' Ames*, Alf Back (coxswain of the QEA launch *Nestor*), Ken Butler*, Harry Dickinson, Frank Finucane, Bert Garrett*, Fred Harriss*, Stuart Hempel*, Lloyd Hill*, George Lake*, Mick Litherland, Mick Pruss*, Arthur 'Speed' Robinson*, Jack Rowe*, Col Sigley*, Alf Thompson*, Dave Thomson*, Bob Timmins* and Fred Tracey. Those indicated with an asterisk are also known to have travelled as flight engineers aboard *Camilla* and *Coriolanus* on the charters to Port Moresby and Milne Bay, in special uniforms supplied to QEA by the RAAF.

One of the first ground engineers to join Eric Kydd in Townsville was Albert 'Sammy' Ames, who transferred from Sydney in June 1942. Sammy's recollection of the period gives some indication of the difficulties under which the QEA engineers were required to work:

> 'It was certainly no picnic. We worked eighteen hours a day, seven days a week with no breaks for "smokos", to keep those aircraft flying. Dave Thomson's wife Rona got a job at the American Canteen and asked if we could get a meal there late at night when we had finished work, because everything else in town was closed. They agreed to this and fed us well, but it was on the proviso that we washed up after our dinner.'

During a particularly difficult period, it was found that the exhaust valves on the Pegasus engines were seizing and needing to be replaced at Townsville after each flight to Port Moresby. It took some time to determine the cause of the problem. QEA had been unable to get supplies of spares for the flying boats from England since the fall of Singapore, so the company engineers began to rework used engine parts. In the case of the exhaust valves, they had a good supply that were slightly worn, and these were honed to 'undersize' and used as replacements. To accommodate the undersized valves, a number of undersized valve-guides were manufactured at the QEA Repair Overhaul Shop at Mascot, and the engines onto which these modified parts were fitted continued to operate without fault. However, a supply of 'standard' size exhaust valves somehow got into the system, and when they were installed into engines with undersized valve-guides and were operated in the high-temperature conditions of New Guinea, they seized-up in the 'open' position.

Each cylinder with a seized valve had to be removed from the engine while the flying boat sat at its mooring out on the water at Townsville. The cylinder was handed to coxswain Alf Back, who would ferry it ashore to Eric Kydd in the workshop. Using a piece of timber and a pair of multi-grips, Eric would eventually free the seized valve, then dismantle the whole cylinder so that he could ream out the valve-guides to 'standard' size. This tiresome process completed, the cylinder was reassembled, then returned to the aircraft and reinstalled on the engine. The whole process was repeated continuously until the problem was completely rectified.

On 22 April 1943 QEA lost another of its flying boats, this time on a regular service flight to Port Moresby. *Camilla*, under the command of Captain Koch, had arrived over Port Moresby harbour in the early evening to find that dark cyclonic cloud had formed and visibility was nil. While waiting for a flare path to be set and conditions to improve, Captain Koch headed the aircraft out over the sea. But his fuel supply was beginning to dwindle, so he decided to attempt a landing at sea despite the adverse conditions. At a height of about 40 to 50 feet, *Camilla* stalled and crashed heavily into the water. Eleven passengers and two crew members were lost. Captain Koch and First Officer Peak survived when they were rescued after swimming for eighteen hours. Records show that no engineering personnel were on board *Camilla* during its final flight.

In July 1943 Captain Bert Hussey and a team of five QEA engineers — Dave Thomson, Sammy Ames, Lloyd Hill, Stuart Hempel and Jack Rowe — flew from Townsville to Horne Island aboard QEA's newly acquired flying boat *Clifton* to carry out transport operations to Tanahmerah, on the Digul River in Dutch New Guinea. The operation, which involved troop and equipment movements, had previously been undertaken by the RAAF's 41 Squadron and was to be flown under the direction of Torres Force. Ten flights into Tanahmerah were planned on consecutive days, and each of the engineers took turns to accompany a flight (in uniform) as flight engineer. Sammy Ames recalls some details of the operation:

'The trips into Tanahmerah were dangerous and because of the risk of enemy action, each flight was accompanied by a fighter escort consisting of two Kittyhawks, one on each wing.

Landing and taking-off from the Digul River was also a hazardous operation. Although the river was shallow, it ran a very strong current and all sorts of debris, including trees, continually floated down-stream.

On one occasion, we had to stand on the wing and wait for trees and debris to clear as our fighter escort circled above. We knew if we waited too long, the escort would run short of fuel and be forced to leave us. We hurriedly disembarked the troops and taking nothing on board, we closed the doors for take-off. The wireless officer and I moved down as far as possible into the tail of the flying boat and stood there hanging onto the aircraft stringers while "Clifton" struggled to become airborne without mishap.

On the ninth trip, the aircraft hit the river bottom as it landed at Tanahmerah. Lloyd Hill was Flight Engineer on that occasion and he lifted the floor boards and inspected the hull and bilge area for damage. Everything appeared to be OK, but in fact the aircraft had actually been holed and we found afterwards that mud and debris had filled and blocked the opening, stopping the water from coming in. As the aircraft took-off though, all the debris was washed away leaving the hole exposed.

When "Clifton" arrived back at Horne Island, the engineers first of all carried out normal service and inspection of the engines and then the aircraft. When the floor boards on the promenade deck were finally raised, we found water about one inch below where we stood, and we realised we were sinking.

We quickly manned the bilge pump, but soon found that this was having little effect and we were fighting a losing battle. The Aldis lamp was flashing a desperate message seeking assistance from shore, and with the help of RAAF personnel and their buckets, we finally overcame the inflow of water and forced rags into the hole.

We took off immediately, but during the take-off the water forced the rags out of the hole and water shot up into the aircraft hitting the metal roof with the force of a fireman's hose, creating quite a racket.

While in the air, we were able to make a temporary repair using a Hull Emergency Repair spare taken from the aircraft's Flight Spares Kit.

As a result of this incident, the proposed tenth flight to Tanahmerah did not eventuate. "Clifton" returned to Townsville where it was found that the temporary repair was still holding and the aircraft was able to proceed to Sydney for major repair.'

Sammy Ames is quite short in stature, and on one occasion in 1944, after a passenger complained that oil had dripped onto his clothing as he alighted from a QEA flying boat, Arthur Baird asked Sammy if he would be prepared to crawl in through the wings behind the engines while the aircraft was in flight, to try to determine the source of the oil leak. When Sammy agreed, Arthur Baird lent him a flying helmet and goggles, and the next morning he was on his way to Townsville. As the aircraft began its flight up the coast from Sydney, Sammy donned overalls, helmet and goggles and crawled up into the wing and out to the port inboard engine, where he lowered the bulkheads and proceed to examine the engine for signs of an oil leak. He remained in that position for about thirty minutes, but finding no evidence of the problem at hand, he replaced the bulkheads and moved out to the port outboard engine and followed the same procedure. Moving back through the aircraft and across to the opposite wing, he carried out starboard engine inspections but was still unable to trace the source of the leakage.

It took two consecutive services to and from Townsville, with Sammy travelling in the wings behind the engines for the entire duration of the flights, to determine that the oil was coming from the engine breather outlet. The 'leak' was considered to be insignificant, and no action was taken to rectify the problem on the engine. Instead, a small metal gutter was fitted above the passenger door to catch any oil dripping from the engines, and no further complaints were received from passengers.

On 22 July 1943 QEA initiative and ingenuity resulted in the reopening of the vital air link between Australia and England. Using modified Catalina aircraft supplied to Britain from the United States under lend-lease, QEA began a top-secret service across the Indian Ocean, flying the 3513 miles from Perth to Ceylon non-stop over enemy-patrolled seas in complete radio silence. The journey took more than twenty-seven hours, and because the service would leave Perth before dawn, fly all through the day and night before alighting on Koggala Lake in Ceylon after daybreak the following morning, passengers were presented with a certificate that entitled them to membership of 'The Secret Order of the Double Sunrise'.

Five Catalinas were made available for the service. All retained their

British civil registrations, but QEA's Captain Bill Crowther (who was Manager of the company's new Western Operations Division) named the aircraft '... each after a star which was used by air-crew in their celestial air navigation'. On the sides of the tail fin they all carried a large white 'fleet number' painted by the British before delivery:

Fleet No.	USN Designation	Registration	Fleet Name	Engine type
1	PBY-5B	G-AGFL	*Vega Star*	2 x 1200 hp Pratt & Whitney R1830-92
2	PBY 5B	G-AGFM	*Altair Star*	2 x 1200 hp Pratt & Whitney R1830-92
3	PBY-5	G-AGID	*Rigel Star*	2 x 1200 hp Pratt & Whitney R1830-92
4	PBY-5	G-AGIE	*Antares Star*	2 x 1200 hp Pratt & Whitney R1830-92
5	PB2B-1	G-AGKS	*Spica Star*	2 x 1200 hp Pratt & Whitney R1830-92

Although all the Catalinas were modified for civil use by BOAC at Hythe before delivery to QEA, they underwent additional modification by QEA engineers at Rose Bay before being reintroduced onto the service. The position of the extra fuel tanks was changed to improve the aircraft's flying attitude, and despite British opposition (for reasons that are unclear), fuel dump valves were installed. The original American engines were replaced by the far more reliable Australian licence-built Pratt and Whitney 1200 hp twin-row Wasp engines, produced by the Commonwealth Aircraft Corporation at the Department of Aircraft Production's Lidcombe annex, where James Kirby (later to become Sir James and a member of the Qantas Board) was in charge.

The QEA base in Perth was at Nedlands, on the Swan River, close to the US Navy Catalina base at Crawley Bay. Norm Roberts was appointed Senior Station Engineer, and Col Sigley his assistant. They had become registered members of the RAAF Reserve and had been sworn in at Rose Bay before departure for the west. They had also been fitted out at Bradfield Park Supply Depot with RAAF uniforms and full flying kits.

Norm Roberts had arrived in Perth with Arthur Baird during June 1943 to set up the engineering operation. They bought a Ford V8 utility, which provided a 'mobile base' for the first few weeks, but as the utility became overloaded with rope and other marine gear for handling the Catalinas, they began to look for storage areas near the river base site. Their search ended when they were given access to four lockable dressing cubicles at the Nedlands swimming baths half a mile away, and although these facilities were less than adequate, the men were grateful to have them while engineering workshops were being set up both at the Nedlands marine base and at Perth's Guildford aerodrome. In a letter to author

Barry Pattison in 1977, Norm Roberts recalled the problems that beset them during what was undoubtedly a period of constant frustration:

> 'Most difficult and destitute circumstances confronted our aim to establish the world's longest regular air service ever then attempted. Difficult because we had to make our start without any facilities for handling, servicing, fuelling, or despatching our aircraft. Destitute because we did not have any servicing equipment or tools, nor any spares except for fifty spark-plugs I had collected in Melbourne on my way from Sydney to Perth. Further destitute because the stubborn opposition to the venture by DCA made it necessary for us to proceed without at least the primary aids and support that should have come from that quarter.
>
> It was a team effort and our determination to succeed that helped us ignore many difficulties we had to overcome during those early months.'

The first two Catalinas to be delivered were *Altair Star* and *Vega Star*, both of which entered service in July 1943. Then in September 1943 the third aircraft *Rigel Star*, arrived, and BOAC asked QEA to extend the route to Karachi. The initial team of six engineers was increased, and an empty kiosk near the base was rented and fitted out with three workbenches, shelving for spares and stocks, and a desk with a telephone. Of the kiosk, Norm Roberts wrote:

> 'It was indeed a great day in mid-October when we moved into the shelter of a building for the first time. We were now in command, for the first time, of our growing work commitment.'

Initially, there were no facilities for refuelling at Nedlands, so the QEA Catalinas had to be water-taxied some two and a half miles to the US Navy base at Crawley Bay to be refuelled.

> 'The existence of an American Catalina base on the Swan River was entirely coincidental, but needless to say it was fortunate for us even though we had been warned not to seek help from that unit.
>
> Our first Catalina G-AGFM ... (although "Altair Star" was given the fleet number "2", it was actually the first to be delivered, on July 7 1943) ... had been delivered to us in very good condition. It had been well cared for by BOAC and efficiently serviced by the RAF. Most of the components, including the engines, were original with adequate working lives still available for our immediate requirements. Our early servicing was therefore directed to functional testing and close examination of the aircraft in general — including of course changing the spark plugs.
>
> We did not have a spark plug spanner. This was a special tool, and so it befell me to ignore the ban and approach the Americans for help. The help I received was perhaps understandably not spontaneous, but I borrowed a spanner on the promise it would be quickly returned. By this means we completed, while afloat, our first servicing routine on our first Catalina.

We continued to borrow the spanner until one day I decided to show my gratitude more realistically to the Crew Chief with whom I had been dealing. Before making my usual request, on this occasion, with a complimentary remark, I presented him with a couple of ebony elephants I had brought back from Ceylon. He looked at these quietly for a moment or two and then he said, with the first smile I had seen on his face, "Man, you have just won yourself a spark plug spanner."

Further exchanges occurred as the need arose, and the dreaded possibility of having to suspend our operations until a vital replacement could be obtained was avoided during our developing or proving period. It might be said that our flying at that time was often sustained by the power of an ebony elephant.

There is no doubt our confidence was further bolstered as each successful operation was followed by another. So it was one morning that Bill Crowther and I were out in mid-stream standing together in our small boat. As we watched our departing Catalina heading out into the nor-western sky, we impulsively turned and warmly shook hands.

It was a spontaneous reaction to the sense of achievement that came from the despatch of our tenth service, and that moment we felt our success was assured.'

Norm Roberts later named the small dinghy they were standing aboard *Twinkle Star* and subsequently had the name, complete with a six-pointed star, painted on the boat. In December 1943 the first Catalina engine change was carried out at Nedlands. But since the DCA had still made no move to construct a slipway at the QEA base, the Catalina had to be towed around to the US Navy base, where permission was given for it to be beached on the condition that the exercise would be brief. The engine was removed quickly and transported to the QEA kiosk, where it was placed on an engine stand. The aircraft was then towed back to its moorings at Nedlands. In the kiosk the engine was stripped and rebuilt — with spares obtained from the Americans at Crawley — and then bench-tested at a well-equipped motor garage in Perth, before being reinstalled onto the aircraft back at the Navy base.

In early 1944 the Shell Company provided refuelling facilities at Nedlands, designed for the Catalinas by Dudley Barker, the Shell Aviation representative for Western Australia. He devised a fuel delivery system that included a jetty to carry the fuel pipeline onto a terminal platform in deep water; a floating extension ramp from the jetty then carried a hose line to the aircraft. The jetty became known as the 'Barker Bridge', and passengers and crew were now able to walk aboard the aircraft in comfort. The first building to be erected for QEA at Nedlands was a long shed, which replaced the kiosk as the engineering workshop and store. From his office in the new building, Norm Roberts designed a nose hangar for Catalina engine maintenance and submitted his proposal to QEA in Sydney. With support from Arthur Baird, finance was approved and the nose hangar eventually built.

But even though plans for construction of the new hangar were well advanced, the QEA base at Nedlands still did not have a slipway. The RAAF had supplied a beaching chassis set for the Catalinas, but slipway construction was the sole responsibility of the DCA. That the DCA remained inactive for so long was nothing short of a disgrace in the circumstances and a total discredit to that Department. As time passed with no action on the slipway being evident, Norm Roberts became increasingly anxious, but his concern was finally eased when:

> '... one day, several truckloads of blue metal and a work team arrived at the base. Work on the slipway commenced and by the end of the day sufficient metal had been deposited and hand-spread to form a nicely graded slope. It now only needed rolling and finishing with a concrete surface and we would be independently established.
>
> After vainly waiting for a couple of weeks for this further work to proceed, I impatiently rang the Works Department only to be told the job was complete. At my request, however, I was promised that an engineer would investigate the problem. He agreed that the ramp of blue metal would be useless for our purposes and explained that, unfortunately, that was all the DCA had provided. He did promise just the same to urgently pursue the matter further.
>
> In desperation, I did make a futile attempt to pull a Catalina up the ramp but almost immediately the wheels [of the beaching dolly] sank into the blue metal and I had to resort to using its engines to get the aircraft back into the water.
>
> The work on the slipway was completed some four to five weeks later, but in the meantime I had temporarily surfaced the ramp with Marsden matting 'borrowed' from the American base. It was by this means we were able to complete 'on shore' inspections of our Catalinas at Nedlands.
>
> The matting remained in our possession, and subsequently it enable us to position aircraft beyond the limits of the tarmac for special work. It thus served on two occasions when sheets of badly corroded bottom plating had to be replaced on G-AGFL [*Vega Star*] and G-AGFM [*Altair Star*].'

Charlie South was initially stationed in Ceylon as Senior QEA Engineer, and he was later replaced by Phil Compston. At Karachi, where the Catalinas usually remained for thirty hours, routine servicing was carried out by BOAC engineers.

Although all regular maintenance was carried out by QEA engineers at Nedlands, major engine overhauls required more extensive facilities than those available at the QEA base. The engines were removed from the aircraft in the QEA nose hangar and packed for rail transportation to Kalgoorlie (some 300 miles inland), where overhaul work was undertaken by the RAAF at Station Boulder. Norm Roberts was obviously pleased with the RAAF's standard of performance:

> 'To the very worthy credit of the RAAF Overhaul Depot at Boulder, we enjoyed the utmost confidence in the dependability of our engines over the Indian Ocean throughout the full period of our Western Operations.'

In 1944 the British Air Ministry released two converted Liberator bombers for use on the Indian Ocean route. They were powered by four 1200 hp Pratt and Whitney R1830 Twin Wasp engines (the same American-built engines that originally came with the Catalinas) and cruised at about 200 miles per hour (70 miles per hour faster than the Catalinas). Separate engineering facilities were set up for the Liberators in Perth at Guildford aerodrome, and a QEA Engineering Station was established at Ratmalana airport in Colombo. Bob Jenkins was appointed QEA's first Station Engineer at Ratmalana, and Charlie Short later became the second.

The first Liberator LB30 (G-AGKT) arrived at Guildford on 3 June 1944 and was placed into service on 17 June. The length of the sea crossing was able to be cut by 436 miles since the Liberator service followed the coastline north from Perth to Learmonth on Exmouth Gulf (on Western Australia's North West Cape), before setting out across the Indian Ocean to Colombo. The route was called 'The Kangaroo Service', and the higher speed of the Liberators reduced the flying time by about ten hours.

Then on 14 August, the second Liberator (G-AGKU) was delivered to QEA, but before it began operation on The Kangaroo Service, it was flown to QEA's workshops at Archerfield, where modifications were carried out to increase its payload. Fuel tanks were removed from the bomb bay, but fuel capacity was maintained by the installation of rubber fuel cells in the mainplane. An additional seven seats were fitted in what had been the bomb bay, and circular port holes with double glazing were located between the landing gear and the entry door for the benefit of passengers travelling in this area of the aircraft. These alterations enabled the Liberator to carry fifteen passengers and a crew of five, giving it a payload of 5500 pounds, five times that of the Catalinas.

Modifications to the Liberator altered its centre of gravity, and because it was essential that the changed position be accurately located, the aircraft had to be weighed. The only place in Australia with equipment large enough to weigh an aircraft the size of a Liberator was the RAAF station at Laverton in Victoria, and so after its test flight at Archerfield, G-AGKU was taken to the southern base. The RAAF officer responsible for weighing the Liberator at Laverton was a young Flight Test Engineer named Ron Yates, who had recently graduated from the University of Sydney as a Bachelor of Aeronautical Engineering. (Ron Yates joined QEA in 1946 and was destined to become the only engineer in the company's history to rise to the position of Chief Executive of Qantas.) G-AGKU returned to Perth and began operation on The Kangaroo Service on 24 September, allowing the original Liberator, G-AGKT, to be ferried to Brisbane for similar modification.

In January 1944 Jan Aldous had employed Phil Ballard as a junior draughtsman and set him up at a desk against the corrugated-iron wall of the Igloo Hangar at Archerfield. Phil was almost seventeen years old and came to QEA straight from school. His desk had a sloped top and full-length drawers to the floor, which meant that he had to sit sideways on a stool as he worked. His drawing equipment consisted of a single T-square. Under Jan Aldous' guidance, he made drawings of small items of equipment that had already been manufactured and repairs that had been carried out, but because of his age and inexperience, he did not initially do any design work.

When the first Liberator arrived at Archerfield from Perth in August 1944, two draughtsmen from the Department of Civil Aviation in Melbourne came to Brisbane to make drawings of the bomb bay modifications. Phil was given the opportunity to assist and made several drawings under their guidance.

On 16 October 1944 QEA's Liberator G-AGKU was coming in to land at Guildford after a test flight, when the nose wheel collapsed, causing major damage to the entire nose section of the aircraft. Norm Roberts flew to Brisbane, where the QEA workshops at Archerfield were repairing and servicing the B-24 Liberators for the US Army Air Corps.

> 'The full section, required to repair G-AGKU, was cut from the nose of a crashed USAAF Liberator being cannibalised [by QEA]. The section was then flown to Guildford where the major rebuilding of the nose was completed in the ANA hangar in four weeks.'

Despite the introduction of the Liberators onto the service in June 1944, the Catalinas continued to operate across the Indian Ocean until July 1945. They made a total of 271 crossings between 1943 and 1945, recording only six in-flight engine shut-downs — a fine tribute to the skill and integrity of workmanship of the QEA engineers.

Three Douglas C47 aircraft were made available to QEA by the Australian Government in August 1944 to maintain a frequent land-plane service between Brisbane and Darwin.They were subsequently converted to DC3 specifications at the QEA workshops at Archerfield, where Pratt and Whitney SC3-G-1830 Twin Wasp engines were fitted, and then placed into service on the new QEA New Guinea route between Sydney and Lae in April 1945, carrying the Australian registrations VH-AFA, VH-AEZ and VH-AEY.

By mid-1945 QEA had more than 1800 employees, a considerable number of whom had served the company in the front-line of enemy activity. Although never officially regarded as 'Service personnel', the QEA engineers, radio operators and pilots who were required to fly as civilians on numerous dangerous missions of enormous strategic and military importance into remote enemy-patrolled areas, did qualify for military service medals and were eventually afforded 'Returned Servicemen' status when hostilities ceased and World War II ended.

CHAPTER

10

PEACE, AND A NEW DIRECTION

Although the war in Europe did not end until 8 May 1945, QEA received confirmation from the Australian Government in March that it was to retain its pre-war role as Australia's overseas airline, and approval was given for a continuation of the arrangements under which the company operated the Sydney–London air route with BOAC. But with the exception of *Coriolanus*, QEA had lost its entire Empire flying boat fleet during the war (*Clifton* and *Coolangatta* were destroyed in separate incidents during 1944 while alighting at Rose Bay on Sydney Harbour); so five converted standard MK III Lancaster bombers were allocated to the company to fly the route.

The Avro 691 'Lancastrians', as the converted aircraft were called, were powered by four 1635 hp Rolls-Royce Merlin XXIV engines (also known as RAF T24-2 engines), each of which was fitted with a single-stage two-speed supercharger. They cruised at 230 miles per hour and carried up to nine passengers. Compared to the luxury of the Empire Flying Boats, however, the cramped conditions on the Lancastrians made them unsuitable as a long-term passenger aircraft.

In late March 1945 senior QEA engineers Jan Aldous and George Williams went to England for a seven-day training program on Merlin engines and the new aircraft. They returned to Australia on the delivery flight of the first Lancastrian, arriving at Mascot aerodrome in Sydney on 17 April.

The first Lancastrian through-service to London left Sydney on 2 June 1945, flying to Learmonth (in Western Australia), then on through Ratmalana (Ceylon), Karachi (Pakistan) and Lydda (Palestine) to Hurn in southern England. Although the Lancastrians flew the entire Kangaroo route to London, QEA crews operated the service only between Sydney and Karachi. The aircraft carried no markings to signify their QEA association and retained their original British civil registrations (G-AGLS, G-AGLT, G-AGLW, G-AGLY and G-AGLZ).

In March 1946 QEA employed its first professional Aeronautical Engineer. Ron Yates was still in the Air Force when he first came to Rose Bay seeking an interview with Arthur Baird. He had been involved in aerodynamic flight test work as a member of the RAAF's No. 1 Aircraft Performance Unit, but now that the war had ended he was hopeful of getting a job with Qantas. He was eventually granted an interview, and the first question Arthur Baird asked was: 'Well, son, what licences do you have?'

Being a professional Aeronautical Engineer, such things as licences never entered Ron's thoughts, so he replied: 'I can drive a motor car. I've got a licence for that.'

Arthur Baird persisted. 'Do you have a licence for a Pegasus or a Merlin?'

Ron was way out of his depth. 'Gee, no,' he replied, 'what's that?'

Arthur Baird just sat there and looked at him for a while and then said, 'I don't think you'd be much use to us here, son. But I'll check with my colleague Lester Brain to see if he can use you.'

Ron went to see Lester Brain at QEA's Head Office at Shell House in the city, and after a lengthy conversation the Operations Manager said to him, 'When can you start? I want you straight away.' (Ron Yates recalls that at the time, 'the Australian Government, through the Department of Civil Aviation, was introducing airworthiness requirements for aircraft. QEA had only operated with aircraft in accordance with British Government licence arrangement up to this period. Aircraft were now being put onto the Australian register and had to meet the latest airworthiness requirements, and nobody in Qantas really understood them or appreciated how to go about getting approval for them.')

Lester Brain rang Arthur Baird to advise that he was going to give Ron Yates a job and wanted him to start as soon as possible. But Arthur Baird was not too happy with that arrangement. 'Now just a minute, Lester,' came the reply. 'He came to see me for a job in the first place. I'll tell you what I'm prepared to do. I'll put him on my staff and second him to you.' So Ron Yates began his long career in Qantas in the Operations Department under Lester Brain, who immediately took steps to secure his release from the RAAF.

It was intended that the converted Lancasters be used to fly the Kangaroo service for a strictly limited period only, because although Britain had concentrated on developing combat aircraft to the exclusion of all else during the war, Great Britain's first post-war airliner, the Avro Tudor, was expected to enter service before the end of 1945. The October 1945 edition of *Qantas Empire Airways* reported that,

> 'The first of the Tudor IIs, of which QEA have several on order, is rapidly nearing completion and should be ready for test flight very soon. The Tudor II is a larger version of the Tudor I, and is intended for the operation of Empire and European services.'

But by December 1945 QEA had discarded the much hoped for Tudor II, which had only just reached its final prototype construction stage, having judged it to be 'uncompetitive, already obsolete and uncertain in performance'. Since it was thought that no suitable British aircraft was likely to be available for at least four years, QEA made a submission to the Department of Civil Aviation, stating that:

> '... serious consideration must be given to the employment of an American type of aircraft ... with the intention to revert in due course to British aircraft. We therefore wish to recommend ... the acquisition of the Lockheed Constellation 649 model for use on the England–Australia route via India, with QEA and BOAC using the same equipment ... fitted with long-range wings, this aircraft has a cruising speed of 255 miles per hour at 10,000 feet, and a payload of 15,800 pounds on stages up to 2,500 miles, after allowing for safety margins. It is fitted with 48 seats by day, convertible to 22 bunks plus four seats by night. The aircraft will be pressurised ...'

BOAC, being a British Government instrumentality, disagreed vehemently with the submission, despite its 50 per cent shareholding in QEA, and continued to lobby for the introduction of Tudor IIs on the England–Australia service. (Interestingly, because of protracted delays with the Tudor I airliner, BOAC had already introduced the 049 Constellation on their 'Speedbird' service across the North Atlantic between London and New York.) But in early September 1946, in a decision that was to change the course of civil aviation in Australia, Prime Minister Ben Chifley announced that QEA had been given authority to purchase four American Lockheed Constellations to operate the Kangaroo service to London — a decision the British Government viewed with great concern.

Then in March 1947 the Australian Government acquired BOAC's shares in QEA, establishing it as a wholly Australian national organisation. But international aviation was now becoming so complex that negotiations for the establishment of airline routes around the world were being conducted on a government-to-government basis, so in June 1947 the Australian Government also purchased the Qantas shares in Qantas Empire Airways Limited. Qantas Empire Airways was now a fully government-owned operation.

The four Constellation aircraft ordered by QEA were the long-range L749 International model, which was pressurised and designed for over-water flying 'where maximum seating was required'. They were powered by four 2500 hp Wright Cyclone R3350-C18BD1 engines and carried thirty-eight daytime passengers and a crew of ten. The Constellations flew at a maximum cruising speed of 318 miles per hour at 20,000 feet and had a maximum non-stop range at top gross weight of 4800 miles. The extra range of the L749s was attained by the inclusion of special fuel tanks built into the outer wing panels, each of which carried 550 gallons of additional fuel. According to Lockheed publicity at the time,

> 'A point of particular interest is brought out by Lockheed engineers in that location of extra fuel in the outer wing panels actually improves the strength of the wing, since the added weight of the fuel counteracts the bending force exerted on the wing in flight by the weight of the fully loaded fuselage.'

One of the features of the Constellation was the provision for carrying the Lockheed 'Speedpak' external cargo carrier, which fitted snugly into the underside of the fuselage and carried 4000 pounds of freight on long-range flights. The Speedpak was loaded separately, then rolled into position under the aircraft and by means of its self-contained electric hoist was raised into position and locked ready for flight. (Accordingly to Ron Yates, the Speedpak concept 'was found to be unsuitable for Qantas operations. It weighed so much when fully loaded it reduced the aircraft's payload range to an unacceptable level, so it was not used very often on QEA flights.')

In preparation for the arrival of the Constellations, a contingent of QEA engineers, under the direction of Doug Hudson, was sent to the United States for training and familiarisation on the new aircraft at the Lockheed plant in Burbank, California, and at the Curtiss-Wright engine factory and propeller plant in Woodbridge, New Jersey, outside New York. (Doug Hudson had joined QEA in April 1939 as assistant to

Works Manager Arthur Baird. He was a University of Sydney graduate with a Bachelor of Science degree in physics and mathematics (1936) and also had a Bachelor of Mechanical and Electrical Engineering (1938). At the outbreak of war he enlisted in the RAAF, where he served full-time between September 1939 and July 1945. Upon his discharge he returned to QEA and was subsequently appointed Superintendent of Technical Development, before being stationed at Lockheed in September 1946 as QEA's first Resident Engineer at the Burbank factory.)

Engineers known to have undergone training on the L749 Constellations in America include Harry Adby, Jan Aldous, Sandy Armstrong, Jack Avery, Len Carolan, Ted Cloherty, Bill Dorrian, Jim Lamb, Arthur Littlejohns, Bob Morton, Brian Murton, Jack Ray, George Roberts, Norm Roberts, Frank Rowley, Bob Scott, Val St. Leon, Dave Thomson, John Upex, Cecil White and Henry Williams.(Val St. Leon was one of few QEA engineers to gain his 'B' commercial pilot's licence and fly the company's aircraft as a Qantas captain.)

By September 1947 the QEA Constellations were beginning to roll out at Burbank. They were given the names and registrations Ross Smith (VH-EAA), Lawrence Hargrave (VH-EAB), Harry Hawker (VH-EAC) and Charles Kingsford-Smith (VH-EAD), and for the first time the 'flying kangaroo' symbol was used as the company's emblem on the aircraft.

Meanwhile, back in Sydney, the groundwork for the reception, servicing and operation of the Constellations was well in hand. Under Arthur Baird (whose title was now Engineering Manager) and Chief Engineer Dudley Wright, renovations to the engineering workshops and extension of the overhaul workshop were completed, and Jack Avery (who had been Senior Station Engineer in Perth since 1945) was appointed Works Superintendent–Mascot. The instrument and electrical shop was updated to cope with the advanced technology of the Constellations and George Roberts placed in charge after his return from the United States.

The first QEA Constellation, Ross Smith, arrived in Sydney on 14 October 1947 following its delivery flight from Burbank. Among those on board was Doug Hudson, returning to Sydney after thirteen months at Lockheed. By the end of the month all four QEA Constellations had arrived from the United States, ready to commence service on the world's longest air route.

In November 1947 Captain Bert Ritchie took the Ross Smith on a proving flight to London. Ron Yates accompanied the flight as Performance Engineer, and his main concern was to specify what payload the aircraft was actually carrying and what loads could be carried. He examined fuel consumption, determined the best cruising altitude, and defined the most economical combination of engine power and speed to get the best range out of the aircraft. This enabled him to set up recommendations to the Flight Operations Department, which resulted in the aircraft being flown in a manner designed to achieve the best results from an operations point of view.

Ron Yates continued to concentrate on the performance aspect of the Constellations and introduced the concept of 'fuel-by-weight' to the company.

> 'In those days when we uplifted fuel, its weight was calculated at 7.2 pounds per imperial gallon, and the Load Chart was based on that standard. Using hydrometers, I conducted a series of 'specific gravity' checks on all take-offs and found that the actual figure was less than seven. As a result, we were able to uplift more weight because the fuel was lighter than we were anticipating by applying the accepted standard, allowing the payload on QEA flights to increase significantly. Although this procedure is standard practice with most airlines today, QEA was the only airline in the world applying it at that stage.'

On 1 December 1947, the QEA Constellation *Charles Kingsford-Smith* left Mascot on the inaugural service to London, with Flight Engineer John Jackson officially listed as a member of the crew. This was the first occasion that such recognition had been afforded the Flight Engineer, and it became his stated responsibility to carry out certain functions that had previously been undertaken by the Captain: synchronisation of the engines and adjustment of the boost to give the best range, speed and most economical fuel consumption. He also kept a constant check on the fuel flow and the amount of fuel remaining in the tanks, as well as controlling the cabin pressure system. When the aircraft was on the ground, the Flight Engineer was responsible for supervising any modifications or checks for unserviceability in liaison with the ground engineers.

By the beginning of 1948 QEA had all four L749 Constellations operating three services fortnightly on the Kangaroo route in parallel partnership with BOAC. Flying time averaged fifty-five hours, and for the first time QEA crews flew the entire 12,137 mile journey to London. But within a fortnight of the commencement of the service the Constellation's temperamental Wright Cyclone engines began to pose problems. Initially it was thought that a spare engine ratio of 50 per cent (that is, eight spare engines) would be sufficient, but it soon became apparent that the ratio needed to be increased to 100 per cent, and sixteen relief engines were located at strategic points along the Kangaroo route, together with an appropriate supply of spare parts. In

Wings to the World Hudson Fysh recalled that the relief engines were:

> '... flown on the Australian end in Liberators, BOAC looking after the UK end, with KLM [Royal Dutch Airlines] doing the overhauls at their renowned overhaul shop at Schipol, perhaps the most advanced overhaul shop of its day. KLM were both efficient and reliable and helped us out of many a difficulty.'

Just before Christmas 1947 the veteran flying boat *Coriolanus* (VH-ABG) came ashore for the last time at Rose Bay in Sydney. It was the final Empire Flying Boat to be retired from service, having flown an incredible two and a half million miles, probably more than any other commercial flying boat in the Empire. Since its delivery to Imperial Airways in 1937 *Coriolanus* had seen an exciting period of war service and recorded just under 20,000 accident-free flying hours. It was broken up at Rose Bay during 1948.

When Lancastrians were withdrawn from passenger services on the Kangaroo route in 1947, QEA purchased four from BOAC for use on their other services, and they were given the Australian registrations VH-EAS, VH-EAT, VH-EAU and VH-EAV. The company introduced three new services towards the end of 1947: Lancastrians began operating a fortnightly service to Norfolk Island and a weekly service under RAAF charter to Japan via Darwin and Manilla; and two reconditioned PB2B-2 Catalina flying boats (VH-EAW and VH-EAX), purchased from the RAAF, began flying charter services to Lord Howe Island.

Because of the totally unexpected number of engine failures being experienced by the Constellations (during the worst period 52 per cent of the engines were removed before their time-expired operating schedule of 800 hours) the problem of transporting spare engines to and from the various line stations along the route had to be addressed. That assignment was given to Ron Yates by Chief Engineer Dudley Wright. He decided to convert one of the company's spare Lancastrians (VH-EAV) into an engine carrier and designed the installation of the Constellation power eggs into the Lancastrian's bomb bay, and also the faring that covered it during flight. The faring was hinged along the starboard side, allowing the spare engines to be loaded and unloaded quickly. Aerodynamic tests and structural analysis of the pod had to be carried out before the DCA would certificate the Lancastrian in its modified form and allow it to re-enter service. This is widely regarded as Qantas' first real feat of engineering, even though the aircraft became fondly known as 'Yates' Pregnant Pup'.

Several test flights were conducted to check the stability of the modified engine carrier, and during these flights it was found that the aircraft had become laterally unstable. To counter this loss of stability the wing flaps were lowered to a ten degree 'out' position; and as subsequent test flights indicated that this adjustment had restored the aircraft's equilibrium, they were locked in that fixation and VH-EAV flew for the rest of its life with its flaps extended ten degrees.

The Lancastrians also suffered from loss of stability when engine trouble was experienced during take-off. Loss of one of the Merlin engines could cause the aircraft to swerve significantly one way or the other, ending up on the soft edge of the runway, where the undercarriage would collapse and be torn off. This happened to VH-EAV on two occasions. In the first mishap, the engine-carrying pod burst on impact, but it was rebuilt, re-installed into the repaired aircraft and returned to service. But on 17 November 1951, the aircraft crashed heavily during take-off at Mascot and was totally destroyed.

Fergus McMaster retired in May 1947 as Chairman of the QEA Board because of ill health, and Hudson Fysh assumed the dual responsibilities of Chairman and Chief Executive. It was a period of rapid growth for the company, and the Board moved to acquire land at the corner of Hunter and Elizabeth streets in Sydney as the site for a future Qantas House.

Following the Government's decision to develop Kingsford-Smith aerodrome at Mascot as the international airport for Sydney, QEA embarked on a general replanning of its engineering facilities. A second hangar, Hangar 85, had been built at Rose Bay towards the end of the war to supplement the original flying boat hangar (completed in 1939). Hangar 85 was dismantled in early 1948 and transferred from Rose Bay to Mascot.

On 1 April 1949 a QEA DC3 (VH-AII) flew the company's final domestic flight from Mount Isa to Brisbane. The following day, Trans-Australia Airlines took over the Queensland and Northern Territory inland services, which Qantas had flown since 1922, and also the operation of the aircraft used for the Flying Doctor Service. QEA would, from this point, operate solely as an overseas airline.

Between February and July 1949, QEA acquired five Douglas DC4 Skymasters from British Commonwealth Pacific Airlines (BCPA), following that airline's decision to introduce Douglas DC6s onto their service across the Pacific to the United States. The DC4s were powered by four 1450 hp Pratt and Whitney R2000-D5 engines and carried a crew of seven and between forty-four and fifty passengers at a cruising speed of 210 miles per hour. They were given the names *Malayan Trader* (VH-EBK), *Hong Kong Trader* (VH-EBL), *Philippine Trader* (VH-EBM), *New Guinea Trader* (VH-EBN) and *Pacific Trader* (VH-EBO) and were placed into service immediately on the company's Pacific Island routes as well as its rapidly expanding operations to the Far East.

By mid-1949 the QEA hangar line at Mascot Aerodrome was working to capacity; Constellations were being maintained in Hangar 85, Lancastrians in Hangar 58, and the new fleet of DC4s in Hangar 20.

It was during this period that one of the few remaining links with the company's origins was severed. Arthur Baird, who by his single-minded diligence had made possible the operation of the infant airlines old and unreliable aircraft, retired as head of QEA Engineering on 30 June 1949. From the beginning he had initiated a very high level of capability and self-sufficiency, which has been maintained to this day throughout the various changes in the company's operations. He always made the Qantas apprentices his personal concern and in 1947, he donated a perpetual shield for presentation to the Dux of the Qantas Apprentice School each year.

The esteem in which he was held by members of the QEA staff and other organisations was shown by the tokens of goodwill he received on his retirement. On Friday 24 June the past and present Directors of the company gave a dinner at which he was presented with a gold and silver cigarette case inscribed with the donors' signatures; a cocktail party was held by the Chairman and Managing Director, Hudson Fysh, on 28 June in the Boardroom; and the following day at Hangar 58 he was presented with a set of golf clubs from the entire QEA staff. Later that afternoon he attended a cocktail party given in his honour by the Shell Company. The staff farewell dinner for Arthur Baird was held at the ballroom of the Hotel Australia, and on this occasion he was presented with a gold wristwatch. Finally the Directors of Tasman Empire Airways (TEAL) held a cocktail party in his honour on 4 July.

Paying tribute to Arthur Baird's twenty-eight years of service to both Qantas and QEA, Hudson Fysh wrote:

> 'I cannot think of anyone who has done more in laying the foundations of Australian Air Transport Engineering, both in regard to the work he did himself and what he initiated; and also in regard to his choice of staff, the character he put into everything he did, and the principles which he imparted to others.'

Arthur Baird was succeeded as Engineering Manager by Dudley Wright, and Doug Hudson moved into the position of Chief Engineer.

Despite QEA's choice of the Lockheed L749 Constellation as the aircraft most suited to its operational requirements on the Kangaroo route to London, the company's 'official' policy was still to give preference to British aircraft. Britain's challenge to the American supremacy in aircraft manufacture came in the form of the De Havilland Comet, a jet-powered aircraft that signalled a new age in commercial aviation. De Havilland had started design work on the Comet as early as 1946. It was powered by four of De Havilland's own Ghost turbine engines, which were almost buried in the aircraft's relatively thin wings, next to the fuselage. Exhaustive weather checks were carried out on the fuselage, and when De Havilland's Chief Test Pilot, John Cunningham, conducted a successful first sustained flight in July 1949, the Comet was considered a triumph. Naturally, BOAC saw the Comet as the answer to its ambitious plans for a worldwide network of jet routes and immediately place orders for fourteen.

But although QEA was interested and had followed the development of the Comet closely, it was not yet ready to place an order because the size, range and economics of the Comet did not match QEA's route requirements. However, the company's Assistant General Manager, C.O. Turner, believed that, from a policy point of view, the development of the Comet was of such importance that QEA could not afford to be only partly informed about it. He recommended that QEA should have their own technical experts on the job consulting with De Havilland. Thus began a long and difficult period in which genuine hope for the success of the new British aircraft would be tempered by cold fact and professional analysis from QEA's growing nucleus of technical and performance engineers.

CHAPTER 11

QANTAS IN NEW GUINEA: A STORY IN ISOLATION

The Australian territories of Papua, New Guinea and New Britain were in a state of turmoil during the period immediately after World War II. The Australian Government's Civil Administration recognised the need for speedy rehabilitation, but shipping had virtually ceased, most of the military forces had been withdrawn, and because of the islands' rugged terrain, roads throughout the territories barely existed. In the opinion of Administrator Colonel J.K. Murray, air transportation provided the only answer.

QEA had been operating its 'Bird of Paradise' service into Port Moresby on behalf of the Australian Government since April 1945 using its Douglas DC3s, and this regular contact with the mainland placed it in a position to provide material assistance to the Administration. Initially this service had been weekly, but when QEA accepted responsibility for all internal air transport services in Papua and New Guinea in 1948, the frequency of flights from Australia was increased to three services per week, and the route was extended to Lae.

Qantas set up headquarters in Lae alongside its war-torn airstrip. Apart from a small hut at the edge of the runway, in which Qantas set up its office administration, there were no other buildings in usable condition. Facilities for engine overhaul were non-existent, and aircraft maintenance was carried out in the open air by QEA's two ground engineers, Keith Vickerman and Joe Taylor. Native labour was used extensively to prepare suitable runways in outlying regions, particularly in the most inaccessible areas of the Highlands, where grass strips were constructed wherever possible, in some instances even on the edge of mountain ranges.

Although many aircraft served in New Guinea in the early years, history has shown that none contributed more to the development of the Highlands than the Qantas DC3 freighters. These modified C47 military freight carriers were flown under what was called 'development' conditions, which allowed them to operate into airstrips that would have been forbidden to them in Australia. DC3s were modified at the QEA workshops at Archerfield before being placed in service in New Guinea. The aircrafts' large cargo doors were left intact, but the cabin interiors had plywood insulation installed, collapsible bucket seats fixed along the walls of both sides, and canvas seats located along the middle for use when up to twenty-four passengers were carried. But even under 'development' conditions the DC3s were unable to fly into a number of airstrips, and for these QEA introduced a fleet of smaller aircraft.

A QEA De Havilland DH84 Dragon (VH-AMN) had been flown from Brisbane to Port Moresby in November 1946 to be used as the Administrator's private means of transport around New Guinea, but it was actually seldom used for this purpose. Instead it began doing freight runs out of Lae. It was soon found that there was more work offering than one aircraft could handle, and in April 1947 a second Dragon (VH-AON) was also placed into service to operate from Lae. By 1950 Qantas had eleven De Havilland Dragons flying scheduled services and charters throughout the territories from Lae, Madang and Port Moresby. Powered by two 130 hp Gypsy Major MK1 engines with two-bladed fixed-pitched propellers, the Dragons had a cruising speed of 114 miles per hour but barely sufficient power to reach the altitudes necessary for flying with a full payload in the Highlands. The criterion for performance in New Guinea, which determined whether or not a Dragon was given a Certificate of Airworthiness, was its ability to climb to 10,000 feet with a maximum of 6500 pounds all-up-weight in less than an hour under test flight conditions.

All QEA Dragons were completely rebuilt at Archerfield before being put to work in New Guinea. The cabin door was moved towards the rear of the aircraft to allow for easy loading of bulk freight, and the wind-driven generator, which provided electrical power, was replaced with an engine-driven power source. Modifications to the cockpit included the installation of an artificial horizon, an essential flying aid for operation in mountainous country.

By the beginning of 1950 the QEA fleet operating the New Guinea internal services consisted of six DC3 freighters (VH-EAP, VH-EAQ, VH-EAR, VH-EBE, VH-EBF and VH-EBG), eleven Australian-built DH84 Dragons (VH-AMN, VH-AON, VH-URD, VH-URV, VH-AEF, VH-AIA, VH-AOR, VH-AOT, VH-AQW, VH-AXL, and VH-BDS), one DH83 Fox Moth (VH-UZD), two PB2B2d Consolidated Vultee Catalinas (Island Chieftain VH-EBC and Island Patrol VH-EBD) and a Grumman Widgeon amphibian (VH-AZO) that had been purchased by the Australian Petroleum Company but was being operated by Qantas.

When Noel Tennant arrived in Lae to start his tour of duty as one of QEA's engineers in late 1949, working conditions were still very primitive. He recalls some of the problems:

> 'We had a small corrugated iron hangar on the southern end of the air strip and although some work was done inside, most had to be done outside in the open. The temperature was constantly in the 40s [Celsius] and at 3 pm each day it invariably rained. Working without cover in the unceasing humid conditions was not easy.'

One of the continuing problems caused by the high temperature and wet weather was the incidence of fungi, which grew inside the stick-and-rag airframes of the DH84s. Noel Tennant again:

> 'This was a major problem with the Dragons. Every four weeks we would have to remove the inspection panel from the fuselage. The job of inspection was usually given to the local native boys, since they were smaller than us. They were told what to look for and if they found any evidence of fungus growth, that whole contaminated section of the aircraft had to be replaced.'

The DH84s were returned to Archerfield when major structural repairs to the fuselage became necessary. Major overhaul of the Dragons' engines and airframes was also carried out at Archerfield each year before the annual inspection for Certificates of Airworthiness.

When Kit Brown replaced Len Donnelly as New Guinea Area Engineer in 1950, there were forty Australian engineering staff stationed at Lae and a number of local native 'engine boys' who assisted with maintenance and freight handling. The engineers only ever had a small number of spare parts to work with and therefore continually needed to improvise, particularly when a problem occurred at one of the many outlying areas serviced by QEA aircraft. On long hauls from Lae (for example, to

Honiara), one of the engineers would accompany the flight. Noel Tennant recalls an incident that occurred at a placed called Yandina in the Solomon Islands:

> 'We had landed in a DC3 and found that the tail wheel was flat. I always carried a spare tail wheel on the aeroplane, as well as other essential spares, but when I went to get the jack, I discovered that it was not there. Here we were, approximately four hours from any recognised maintenance base, with a DC3, a flat tyre and no jack to change the wheel. I had no choice but to improvise, and this came in the form of twenty-five natives who stood with their backs under the tailplane, which they then proceeded to lift high enough for me to pull the axle out and replace the wheel.'

Although the QEA Dragons operated successfully in New Guinea between 1946 and 1953, their age was against them, because they had been built in the early 1930s. A replacement aircraft was sought, and the De Havilland Drover seemed to be the most suitable (on paper, at least). The Drover was a robust all-metal monoplane with a fixed undercarriage, simple wing flaps and variable-pitch electrically controlled propellers, driven by three 145 hp Gypsy Major 10 MK2 engines. It was built by De Havilland Aircraft Company Pty Ltd in Sydney and designed for durability and safe performance after the loss of an engine in flight. QEA ordered seven Drovers and took delivery of the first at Sydney's Bankstown airport on 13 September 1950.

By July 1952 QEA had received and placed in service five of the seven Drovers (VH-EBQ, VH-EBR, VH-EBS, VH-EBT and VH-EAZ). But the Drover was proving totally unsatisfactory for operation in New Guinea — apart from the fact that its rate-of-climb performance lagged, it developed chronic propeller trouble, resulting in the loss of one aircraft (VH-EBQ) in July 1951 when it crashed into the sea near Lae, killing all six passengers and the pilot. Several near-misses with QEA's other Drovers were also recorded when the aircraft suffered loss of control because of structural failure of the centre propeller during flight.

Another problem with the Drovers was the position of the pilot's seat. It was located so far back from the windscreen that vision forward and to the right was quite inadequate, particularly in bad weather. QEA engineers in Sydney replaced the original curved windscreen with flat panes, and although this modification improved pilot visibility, the company became so dissatisfied with the Drover that, after delivery of the fifth aircraft, it cancelled its order for the final two.

Meanwhile, at the company's headquarters in Lae, QEA had built a modern air terminal with facilities for office administration and passenger handling, and a hangar for aircraft servicing and maintenance. Using pre-cut timber transported by DC3 freighters from mills at Bulolo in the Highlands, QEA also constructed an autonomous village with all amenities to provide first-class accommodation for passengers as well as airline personnel.

By the end of 1952, the four-engined Douglas DC4 Skymasters had replaced the DC3s on the 'Bird of Paradise' service from the mainland, and QEA had fourteen aircraft controlled from Lae, Port Moresby and Madang, flying to 103 points on regular services and charters. The two Catalinas were flying to Rabaul and the Solomon Islands, completing a network of internal scheduled services which operated an unduplicated total route distance of 5000 miles. They also opened up territory that would otherwise have been impossible to service.

The Catalinas were based at Port Moresby, but when major airframe overhaul was necessary, mainly because of corrosion, they were flown to Lae. Beaching the Catalinas at Lae was a major undertaking, requiring a large amount of physical labour. Because the aerodrome was not equipped for flying boat operations, the aircraft had to be brought ashore at a graded section of the foreshore where Marsden matting was laid over the stony surface. Native boys would swim out to the aircraft and attach the beaching gear before the aircraft was literally pulled out of the water by up to forty natives and QEA engineering staff. (The Marsden matting was then taken up and stored away until required again.) A tractor towed the Catalina through the jungle to a road that ran beside the aerodrome, then along the road and up a ramp onto the airstrip and down the runway to the QEA hangar. On arrival at the hangar the Catalina was inspected for corrosion and as many sheet-metal repairs as could be made in the time available were carried out, with particular concentration on areas below the water line.

Doug Stepto was a fifth-year apprentice when he was transferred from Sydney to the QEA engineering section at Lae in 1956. He recalls the difficulties involved in getting the Catalinas back into the water after overhaul:

> 'The airstrip at Lae was a good deal higher than the level of the Gulf and to get the "Cat" off the strip, you went down a fairly steep incline and turned 90 degrees onto the road, before moving through the jungle to the launching area.
>
> To tow the "Cats", a piece of waterpipe with a set of shackles at each end was attached to the aeroplane and the tractor. The idea was to keep the aircraft perfectly in line with the tractor while it was moving, since the beaching gear attached to the aircraft had no brakes. To stop, the native boys (who would walk alongside the aircraft while it was being towed back to the water) would simply put blocks under the wheels of the beaching gear.
>
> On one occasion I was driving the tractor, and as I began to descend the embankment down to the road, the "Cat" lost its line behind the towing vehicle and began to overtake it, drawing level with the tractor as we careered out of control

across the road, coming to rest in the bush opposite.

With bits of branches protruding from the wing tips and floats, we eventually got going again and proceeded down the road with the natives pulling the foliage away. We were lucky that the aircraft was not damaged in the incident, but my main concern was what Kit Brown would have to say when we finally arrived at the launching area.'

QEA's choice of aircraft to replace its unwanted Drovers was the DHC2 Beaver, a versatile all-metal monoplane built by De Havilland in Canada. It was designed to meet the Canadian 'bush' pilots' dream of an ideal aircraft and seemed perfect for QEA's operational requirements in New Guinea. The company ordered four Beavers and put them into service during 1954 and 1955 (VH-EAS, VH-EAT, VH-EAU and VH-EAV). Powered by a single 450 hp Pratt and Whitney R-985 SB3 nine-cylinder, single-row, air-cooled radial Wasp Junior engine, the Beaver was easy to operate and proved to be very reliable, cruising at 143 miles per hour carrying a pilot and six passengers. It was able to achieve a fast initial rate of climb of 1020 feet per minute, and its low landing speed and stout undercarriage enabled it to handle short, rough runways with ease. The Beaver could also be fitted with floats giving it amphibian capabilities of landing both on water and land. In a review of the Beaver written for the December 1954 edition of *Aircraft* magazine, Ron Yates reported:

'The full power of the engine (450 BHP) is available from sea level up to 5000 feet altitude, which makes it very attractive for New Guinea at such places as Bena Bena (5500 feet), Kerowagi (5210 feet) and Mount Hagen (5500 feet). The supercharger is single-stage centrifugal direct driven.

The fuel is carried in three tanks in separate compartments suspended from the cabin floor. The tanks are metal, two being of 29 gallons capacity and one of 19 gallons. For long-range operations an external metal tank of 36 gallon capacity can be strapped under the fuselage slightly to the rear of the pilot.

The Bendix Stromberg Carburettor NA-R9-C2-6 is fitted with an automatic and manual control for the fuel/air ratio. The mixture control lever is located at the right side of the control quadrant at the top of the pedestal and provides Auto Lean, Auto Rich, Full Rich, Full Lean and Idle Cut-off. Carburettor heat is provided by ducting some of the intake air through a heat exchanger muff which surrounds a section of the engine exhaust collector.

The propeller is a Canadian Pratt and Whitney (Hamilton Standard under licence) two-bladed metal, constant speed unit of 8 feet 6 inches diameter. A pitch control lever is located on the quadrant on top of the pedestal and provides Course Pitch, Decrease RPM and Increase RPM. The governor hydraulically moves the propeller blades to lower angles by engine oil supplied by an engine-driven pump;

when the governor allows oil to drain from the propeller, counterweights progressively move the blades to higher angles.'

In 1956 one of the QEA Beavers featured in an important role in the Australian feature film Walk Into Paradise, which was made in colour on location in New Guinea and starred Chips Rafferty. It was also the first Australian feature film to be entered in the prestigious Cannes Film Festival.

When six young New Guinea boys began working with QEA at Lae in December 1956, they were the first local apprentices to be employed in the aircraft industry in the Territory. Their names were Malai, Kamah, Robert, Simon, August and Joab, and they had all received primary school education at the Malaguna Native Training Centre in Rabaul and secondary schooling at the Lae Native Technical School. Four of the boys were apprenticed as aircraft mechanics' assistants, one as a motor mechanic and one as a painter/decorator, and after three months' probation they entered their first-year apprenticeship in April 1957. They spent six months in each section of the engineering operation, working in such areas as aircraft overhaul, engine build-up, electrical overhaul and servicing, and sheet-metal work. A monthly report on the boys' progress, conduct and ability was prepared by each section head and a six-monthly report submitted to the Native Apprenticeship Board. Their secondary schooling at the Lae Native Technical School included English, Maths, Basic Fitting, Principals of the Combustion Engine, and Aircraft Maintenance Theory. When they were on the job at the QEA hangar, they wore a special uniform of blue shorts and shirts to distinguish them from the other local native employees. QEA's Area Manager for New Guinea, Captain Orme Denny, wrote in 1958 that 'Qantas is proud to be part of this vital phase of the New Guinea development programme'.

In 1957 Jack Prince took over from Kit Brown as Area Engineer in charge of the company's New Guinea operation, and the following year QEA replaced its four Beavers and two Catalinas with four DHC3 Otters, which, like the Beavers, had also been built by De Havilland in Canada. (The final scheduled flight of the QEA Catalina Island Patrol ended an important chapter in Qantas history. When it alighted on Port Moresby harbour on 28 July 1958, twenty years of Qantas involvement with flying boats came to an end. Both Island Chieftain and Island Patrol were subsequently reduced to spares in November 1958.)

The Otters were larger and had a greater range than the Beavers and could also operate as amphibians but at a decreased payload and performance. Jim Burnett, QEA's Assistant Works Foreman at Lae and Ground Engineer Ron Leeds from Port Moresby, came to Sydney in early

May 1958 for training on the new aircraft type at De Havilland's factory at Bankstown. Later that month the first of the Otters (VH-EAW) was handed over to QEA's General Manager, C.O. Turner, at Bankstown aerodrome by the man who was Qantas' first non-wartime pilot and who had become De Havilland's Managing Director, Lester Brain. On 4 June, at a ceremony at Port Moresby, Mrs Shane Paltridge, wife of the Minister for Civil Aviation, christened it *Kikori*. The other three were subsequently named *Kokopo* (VH-EAX), *Kieta* (VH-EAY) and *Kerowagi* (VH-EAZ).

Powered by a single 600 hp Pratt and Whitney R-1340 radial engine, the Otter carried ten passengers and a crew of two. Despite a reduced rate of climb, which was a definite handicap in New Guinea conditions, the Otters were still considered a better proposition than the Beavers, and they flew reliably and without serious incident.

By 1960 QEA had been operating in New Guinea for fourteen hectic years and had made a significant contribution to the rehabilitation of the Territories. Its engineering staff, working in sometimes untenable conditions, maintained the company's standards of excellence and efficiency in servicing and overhauling engines and airframes. Then, on 1 September 1960, QEA withdrew from New Guinea at the direction of the Australian Government, which wanted the company to concentrate solely on its rapidly expanding international routes. The New Guinea operation was taken over by Trans-Australia Airlines. For Nanandai, a Papuan who had been employed by Qantas for many years as an engineering foreman, it was an emotional moment as he lowered the QEA flag at Lae airport for the last time. Within minutes of the company's emblem being furled, the first TAA 'Sunbird' service in the Territory took off for Madang.

Noel Tennant:

'Working as an engineer in New Guinea was undoubtedly the greatest experience I ever had — an experience the extent of which the young engineers of today could not begin to appreciate. We worked by ourselves for a great deal of the time and we had to learn to improvise. We were unable to ring up Production Planning or some other department with a specific problem; we had no alternative but to go out and fix it ourselves.'

CHAPTER

12

THE INFLUENCE OF PROFESSIONALS

After his many perilous exploits on company service in forward areas of enemy activity during World War II, Bill Bennett opted to return to his pre-war posting as soon as hostilities in the Pacific region ceased. He travelled aboard the first flying boat back into Singapore after the war and remained as QEA Station Engineer for just over four years. Then in 1950 he returned to Sydney and became Works Superintendent–Engineering at Rose Bay.

The QEA flying boat base at Rose Bay remained in operation after the end of the war despite the fact that the company had lost all but one of its Empire Flying Boats (*Coriolanus*) in wartime service for the Australian Government. Ern Aldis had become Works Superintendent at Rose Bay in early 1946, just as QEA began supplementing its Lancastrian services on the Kangaroo route between Sydney and Singapore with eighteen of BOAC's British-registered Short S25 Hythe-class Flying Boats.

The Hythe Flying Boats were powered by four 1030 hp Bristol Pegasus 38 engines (some were Pegasus 48 engines) and carried up to twenty-four passengers and a crew of seven at an approximate cruising speed of 170 miles per hour.

Prior to this, Eric Kydd had returned to Rose Bay from Townsville in 1944 as Works Foreman in charge of the conversion program of the PBY Catalinas to operate the Indian Ocean service. When Ern Aldis transferred to Head Office at Shell House as Engineering Administration Officer in January 1948, Eric Kydd was chosen to replace him as Works Superintendent at Rose Bay, a position he held until late 1949, when he was assigned to Mascot and promoted to Assistant Chief Maintenance Engineer.

In the early 1950s Qantas purchased five Short S25 Sandringham Flying Boats, a civil version of the legendary wartime Sunderland 5s. Bigger, faster and more modern than the Empire C-class Flying Boats, the QEA double-decker Sandringhams were powered by four 1200 hp Pratt and Whitney R1830 Twin Wasp engines. They carried up to thirty passengers and crew of seven but at a reduced cruising speed of 145 miles per hour.

(Tasman Empire Airways Limited (TEAL) had replaced its Empire S23 Flying Boats with S25 Sandringhams on the Auckland–Sydney service as early as 1946, naming them 'Tasman' class Flying Boats; they were serviced in Sydney for TEAL by QEA engineers at Rose Bay and operated efficiently on the trans-Tasman run until 1948, when overheating of the Pratt and Whitney engines caused their entire fleet to be grounded for five months.)

The first two QEA Sandringhams (VH-EBX, named Pacific Chieftain, and VH-EBW unnamed) were bought from TEAL in April 1950, and after modification at the QEA workshops at Rose Bay they began operating on the company's new services to Fiji and New Hebrides.

To augment these services, a further three S25s were purchased from BOAC, two of which (VH-EBV, named Pacific Warrior, and VH-EBZ, named Pacific Explorer) were ready to be handed over to QEA by July 1951. It was decided that a QEA crew under the command of Captain Hugh Birch should go to the United Kingdom and make the delivery flight to Australia. Bill Willcox went along as Engineer Officer, and Ivan Hewitt and Jack Kerr, two experienced engineers from Rose Bay, were also sent to accompany the flight.

Since flying boats were no longer operating between Great Britain and Australia, much of the time in England was spent planning the trip back to Australia; when the route was chosen, arrangements for refuelling had to be organised. After final test and acceptance flights had been made, the first S25 (VH-EBV) took-off from Southampton Water at 10 pm on 4 July 1951, followed closely by the second (VH-EBZ), which was being flown to Australia by a BOAC crew. The flight plan took the aircraft through

Marsa Xlokk (on the east coast of Malta) to the RAF base at Fanara (on the Great Bitter Lake at the southern end of the Suez Canal), then over the Arabian Desert to Bahrain, Karachi, China Bay in Ceylon, and across the Bay of Bengal to Singapore, where both aircraft were given a complete check by the QEA engineers stationed at Seletar Aerodrome. The final leg of the journey took them through Sourabaya, Darwin, Cairns, and on to Sydney, where they alighted at Rose Bay on 15 July.

The fifth QEA Sandringham (VH-EBY, named Pacific Voyager) was delivered, modified and placed into service during December 1951.

On 10 June 1951 VH-EBW struck a coral reef while taxiing prior to take-off at Vila, in the New Hebrides, and was holed below the waterline at several places between the keel and chine in Compartment 'A'. It sank quickly and was completely lost.

QEA began a regular fortnightly Sandringham Flying Boat service from Rose Bay to Port Moresby in July 1952 to supplement the Catalina operations on the New Guinea internal routes. The aircraft flew from Sydney via Brisbane to Hayman Island, where it remained overnight before proceeding to Port Moresby. Because fuel supplies were limited at Hayman Island, an additional refuelling stop was also made at Cairns. On arrival at Port Moresby the Sandringham began a ten-day series of flights on the Catalina routes down through the Solomon Islands, before returning to the QEA Flying Boat Base in Sydney. The larger, more comfortable aircraft was reported as being a welcome addition to QEA's New Guinea services.

The four QEA Sandringhams continued in operation until December 1954 when QEA announced the closure of its Flying Boat Base at Rose Bay. Pacific Chieftain was sold to Ansett Airways, and the other three were subsequently reduced to spares at Rose Bay during 1955. The Sandringhams were replaced on QEA's Pacific Island routes by the company's DC4 Skymasters.

On 7 September 1949, BOAC's fleet of DC4B Canadair Argonaut-class aircraft, which had recently replaced flying boats on their Far Eastern routes, were grounded after problems with the glycol pump seal in the Rolls-Royce Griffin engine's second-stage supercharger intercooling system. QEA was asked to fill in for BOAC by operating these services, and under the supervision of Assistant Operations Manager Bert Ritchie, Qantas established a base of operations at Hong Kong's Kai Tak aerodrome. Three of the company's DC4 Skymasters were allocated to fly the weekly BOAC Karachi–Tokyo and Karachi–Manila routes and twice-weekly Singapore–Hong Kong route. Although a major undertaking for QEA, requiring thirty-seven staff to be relocated from Sydney, the whole Hong Kong operation lasted just six weeks. Assistant Chief Maintenance Engineer Eric Kydd headed the engineering team from Mascot, which comprised Keith Ower, Gordon Hodge, Lloyd Hill and Bob Jenkins.

BOAC had no maintenance facilities at Hong Kong. Servicing of their aircraft was carried out by Jardine Aircraft Manufacturing Co. Ltd, and their ground engineers were also contracted to maintain QEA's DC4s while in Hong Kong. The QEA engineers flew aboard the DC4s as Engineering Officers, their prime responsibility being to supervise servicing procedures while aircraft were on the ground at various stopping points along each route.

On 16 September Eric Kydd and Keith Ower arrived at Kai Tak aerodrome as the QEA Skymaster VH-EBM was being made ready for flight. The aircraft stood on the tarmac up against wooden chocks. Cargo was being loaded from a timber stand, and all four engines were being run up. While this was happening, one of the contract engineers released the parking brake. The aircraft immediately jumped the starboard chock and began swinging to port. The propeller of No. 3 engine struck and splintered the cargo loading platform, sending jagged pieces of timber through the fuselage in the vicinity of the cargo door, killing one of the Chinese cargo handlers and severing the aircraft's overload oil reserve tank. Damage to the fuselage was not only extensive but serious, and Lloyd Hill had to remain in Hong Kong as QEA's Engineering Representative and Inspector while repairs to VH-EBM were carried out by Jardine Aircraft Manufacturing.

The Qantas Apprenticeship Scheme had been regularised in 1945 immediately after the war when its Technical Training School was established in the Engineering Department and placed under the supervision of a Master of Apprentices and an Apprenticeship Advisory Committee within the company. By 1951 the Technical Training School had expanded from a single office with one instructor to a large well-equipped educational facility in Mascot's Hangar 20, with four soundproof lecture halls, a private study and six specialist instructors. It was recognised by the Department of Civil Aviation as the only approved airline training school with licenced instructors in Australia. Branch schools also operated in Brisbane and Singapore, and correspondence courses were provided for line station personnel. The aim of the technical school was to ensure that the company maintained a steady supply of fully trained and licenced engineers to service all areas of its developing operation. By mid-1951 more than 300 apprentices had passed through the school, and 715 ground engineers' licences and endorsements had been obtained.

Qantas employed between twenty-five and thirty engineering apprentices each year. They had to have a standard of education equivalent to Intermediate Certificate level and show a marked interest and aptitude, not only in engineering, but in civil aviation as a whole.

After a three-month probationary period, indentures of apprenticeship were drawn up for a normal course of five years' training. For the first two and a half years the apprentices were transferred through most sections of the Engineering Department in the company workshops at Mascot and Rose Bay. Queensland apprentices trained at Archerfield and Moorooka. At the end of this initial training period, each individual was given the opportunity to specialise on engines, airframes, sheet metal, instruments or electrical equipment, and the next two and a half years were spent in concentrated training in their chosen specialised field.

In addition to the theoretical knowledge acquired at the Qantas Technical Training School and practical experience gained on the floor of the QEA workshops, the company apprentices were also expected to undergo further instruction at Sydney or Brisbane Technical Colleges in evening courses that expanded their understanding of their chosen field of specialisation.

The first chairman of the Apprenticeship Advisory Committee was Doug Hudson, who had recently been appointed Superintendent–Technical Development. He headed the Committee from 1945 to November 1953, when Ern Aldis (Chief Maintenance Engineer – Mascot Hangars) took over as Chairman. Eric Kydd (Superintendent Engineering – Mascot Hangars) became Vice-Chairman and Fred Caterson (Assistant Staff Controller) was appointed Secretary. The other members of the committee were Alan Cropper (Senior Staff Officer), Len Carolan (Technical Training Supervisor) and Don Parmee (Master of Apprentices).

When World War II ended in 1945 QEA closed the engine overhaul shop it had been operating for the Department of Aircraft Production at Randwick, and the site was occupied by the company's Stores Accounting Section. It was also used extensively as an area for storage. Then in 1949 QEA re-established a workshop facility at Randwick to manufacture and test small components and engineering equipment to be used at the company's main base of operation at Mascot aerodrome. Dave Thomson, who had been QEA's Station Engineer in Darwin since 1945, was appointed A/Equipment Production Engineer in charge of Randwick. The workshops employed eighty engineering personnel and were split up into two sections: a machine shop and a fitting shop, in which rework, welding, plating, painting, sheet metal, overhaul of graviner fire extinguishers (which Qantas had taken over from the agents, Allison Gary and Co.), and general maintenance functions were carried out. By 1956 the engineering staff at Randwick had increased to 110.

On 1 July 1952 QEA introduced its 'Staff Suggestion Scheme' to encourage staff to submit ideas for improving efficiency and cutting costs within the company. Eighty-six suggestions were received, and the first major award went to Spencer Sweetman, a fitter and turner from the

Randwick workshop. He was presented with a cheque for £52 ($104) by the General Manager, C.O. Turner, at a small gathering of senior executives in the Shell House conference room. The award was for his design and installation of a jig for rework of split ball joints used in Lord Dynafocal engine mount assemblies. (Before incorporation of this jig, it was necessary to fit new ball joints and other associated parts at each overhaul of the assembly, at considerable cost to the company.)

In February 1953 another Randwick employee, Workshop Fitter Gordon Carruthers, became the first QEA employee to received the maximum award in the Staff Suggestion Scheme, a cheque for £100 ($200). His award was for a salvage and repair plan for exhaust assemblies used on the DC4 Skymasters' Pratt and Whitney R2000-D5 engines. The exhaust segment joint assemblies on the Pratt and Whitney engines had a reliable lifespan of only 1000 hours, after which the assemblies would work loose and allow exhaust gases to escape; and because there was no suitable method for their repair, they were scrapped. In an interview given at the time of his award, Gordon Carruthers explained how he developed his scheme:

> 'One day I was short of a segment joint to re-assemble an exhaust ring for a Pratt and Whitney engine and decided to look on the scrap heap. Because there were so many old joints on the heap, I resolved to try to do something about it.
>
> After a lot of experimenting, I evolved a tooling set up whereby I could repress the joints. It was not long before we had every joint on the scrap heap repressed and ready for another one thousand hours.'

According to Dave Thomson (whose position was now Randwick Workshops Superintendent), 'Carruthers' method solved the exhaust segment problem and would save the company a considerable amount of money.'

These were typical of many cost- and labour-saving initiatives developed by engineering personnel over the years, a good number of which were adopted by the Engineering Status Committee and put into operation in the company's workshops.

In July 1951, the QEA Instrument and Overhaul Shop at Mascot was awarded three major government contracts: for the overhaul of Hall Scott marine engines for the RAN; overhaul and flow check of carburettors for the RAAF; and overhaul of aircraft instruments for both the RAAF and the RAN. The contracts were negotiated through the District Contracts Board for the naval work, and the Department of Aircraft Production for the RAAF.

The Hall Scott program was for an initial period of twelve months at the rate of one engine per month, at an estimated 800 manhours per engine.

QEA's engineers completed seventeen engines in thirteen months, all within the number of hours allowed, resulting in the contract being renewed for a further twelve months. The carburettor project was for 120 overhauls and flow checks, and it was felt at the time that QEA was given this order because its carburettor shop had an American-manufactured Stromberg Flow Bench available and in use at Mascot. The government contract work on instruments had been a QEA commitment for several years, constituting about 20 per cent of the Instrument Shop's output and equalling 170 instruments per month. Renewal of this contract was a further indication of the RAAF's confidence in the integrity of workmanship of QEA's instrument engineers.

In the late 1940s the Federal Government established a statutory authority to implement a scheme to dam and divert the waters of the Snowy River, in the Southern Alps of New South Wales, westward into the Murray and Murrumbidgee rivers, so that the dry plains west of the Great Dividing Range could be irrigated. As the project came close to reality the engineers conceived that it would also be possible to generate large quantities of peak-load electric power for New South Wales, Victoria and the Australian Capital Territory. Thus began construction, under the auspices of the Snowy Mountains Hydro-Electric Authority, of what is regarded today as one of the great engineering wonders of the world.

Qantas engineers became involved in the Snowy Mountains scheme in the early days of the initial contour survey, when the Authority's engineers were attempting to determine what levels the water would reach in the hills and valleys behind the proposed dam sites. Because of the ruggedness of the mountain terrain and the thickness of the bush in which surveys had to be carried out, traditional methods of horizontal and vertical angle measurement using theodolites were found to be impossible, so the surveyors came to George Roberts at the QEA Instrument Shop at Double Bay for help. As George Roberts describes the situation:

> 'I was able to offer a simple solution to the problem. I suggested that the S.M.A. engineers carry sensitive aircraft altimeters with them into the areas to be surveyed. But before this could happen, the survey team would have to be trained, and QEA's offer of instruction in the use and care of the small, delicate instruments was readily accepted.
>
> We also recommended that the engineers carry 'aneroid' barometers on their surveys into the mountains. These instruments would record changes in barometric pressure — changes which would affect the accuracy of the altimeters. Without them, the surveyors would not be able to make the necessary sensitivity adjustments to the altimeters and achieve the accuracy required. Two-way radio communication with the base, where simultaneous readings were taken from a 'mercury' barometer, gave further accuracy to the contour surveys, well within the accepted tolerances for an

initial mapping of the area.'

The altimeters were supplied from RAAF stock and were overhauled at the QEA Instrument Shop at Double Bay before being used on the survey. The 'Aneroid' and 'mercury' barometers, together with protective carrying cases and shoulder straps, came from A.L. Franklin, a scientific instrument maker in Sydney, with whom QEA had a close association and good working relationship. QEA's contribution to the success of the survey operation was later acknowledged when the Snowy Mountains Hydro-Electric Authority wrote to Qantas expressing its grateful appreciation of the company's instrument engineers.

Prior to the introduction of the L749 Constellation in 1947, Qantas had mainly operated aircraft because they were available and not necessarily because they were the best aeroplanes for the job. Now it was selecting its aircraft types carefully and purchasing only the most suitable to fly its rapidly expanding network of international air routes. In September 1951 Hudson Fysh wrote:

> 'The standard of any air service is first based upon its equipment. The 'Constellation 749A' has proved a magnificent aeroplane. It has been the main factor in the high standard attained by the Sydney–London service operated by Qantas Empire Airways and British Overseas Airways Corporation.
>
> The '749A' has already been in service for over three years. Six years is regarded as a full obsolence period for major trunk-route aircraft. As aircraft cannot be bought off the shelf, but must be ordered some years ahead, the matter of re-equipment is naturally a vital question for Qantas as Australia's overseas operator. In addition, Qantas must have new aircraft for new routes authorised.
>
> In selecting aircraft for international operation, the route requirements are a first consideration. On the Sydney–London route, via India (the Kangaroo Route), where a heavy mail and freight traffic exists, in addition to passengers; where future landing points could be problematical, owing to threatened internal disturbances, it is clear that long-range and large volumetric capacity aircraft are required. Economy of operation is also most important.
>
> The critical position of fuel supplies in the middle section of the Kangaroo Route is a case in point, and long-range aircraft have a chance of using alternative routes and refuelling points and avoiding critical stops.'

Although the Comet 1 was not considered to be up to Qantas' requirements in such vital factors as payload-range and competitive economic structure, some interest was shown for a time in the Rolls-Royce Avon engined Comet 2 as a possible replacement for the QEA Constellations. However, in the view of the Qantas engineers stationed in Britain to monitor the project, the Comet 2 would remain experimental for a number of years. Delivery to QEA would be 1955 at the earliest.

Another British option available to Qantas was the Bristol 175 turbo-prop airliner the Britannia, which was to be powered by four Bristol Proteus turbo-prop engines. But since the prototype of the Britannia would not be completed before 1952 and delivery of it could not be expected before 1955, Qantas began to look elsewhere.

The two main types considered were the American Lockheed Constellation 1049C and the Douglas DC6B. On the figures available to Qantas, the Constellation 1049C — the Super Constellation — appeared to be the best aircraft to fulfil the company's route requirements. It was recommended for its competitiveness, its compatibility with QEA's existing fleet of Constellation L749s, and the fact that it was to be fitted with Curtiss-Wright Turbo-Cyclone 18 compound engines, which had been developed from the Truculent Turtle engines that were to be installed in the Lockheed Neptune bombers ordered by the RAAF.

Despite fierce lobbying from the British Government urging a decision in favour of the Comet 2, the Australian Government gave approval for QEA to purchase the Super Constellation, and in February 1952 Qantas appointed Ron Yates as its Resident Engineer at the Lockheed factory in Burbank. Ron Yates explains the function and responsibilities of the QEA resident engineer:

> 'Contractually, when you sign up to purchase an aircraft, you not only have a specification, terms and conditions, but also the provision to have a Technical Representative in residence at the factory. From the manufacturer's point of view, it is important to have a voice representing the buyer on hand so that decisions can be taken quickly when changes occur. The aircraft is usually in the development stage when the contract to purchase is first signed, so a great number of factors will invariably change during the course of the programme.
>
> The task of the Resident Engineer–Burbank was to convey to his management back in Australia details of the types of options available, the exact nature of decisions which needed to be made and by when those decisions were required, so that the manufacturer could keep his production line on schedule in accordance with the programme.
>
> My specific task was to understand the aircraft itself. I did courses on the aircraft and also on the engines and components. By the time I was finished, I knew that Super Constellation backwards.
>
> It was also my responsibility to maintain communication with the representatives of other buyers (e.g. TWA, Pan-Am, etc.) to see what they were doing with their aircraft. It was important that we remain competitive, so if I found that they had made some modification in the manufacture of their aircraft that could be applied to advantage in Qantas operations, it was my job to ensure that my management were advised of the details of the alterations and that we were given the same option to have

our aircraft manufactured in the same way.'

In 1951 C.O. Turner succeeded H.H. Harman as General Manager of QEA. Captain Bert Ritchie (who was to become Chief Executive of Qantas) was later moved to comment: 'The grand design in Qantas for many years, certainly in the important years, was the brainchild of Cedric Oban Turner. That is the only way to state it.' Hudson Fysh continued as Chairman and Managing Director.

Qantas aircraft were given a new colour scheme in 1952. The top surface of the fuselage was painted white, the thin blue line below the cabin windows was replaced with a dark red band, and the name Qantas Empire Airways was changed to read simply 'QANTAS'.

In 1948 Ron Yates had advertised for and subsequently appointed two more University of Sydney graduates to work as performance engineers at Qantas. Bob Walker, awarded Bachelor of Engineering (Aeronautical) in 1945, had been working as a cadet engineer doing flight testing work on the new DHA3 Drover aircraft at De Havilland's Bankstown facility. He joined Ron Yates' team at Qantas in early 1949 and was followed a week later by Bill Hill. Although they had graduated in different years, all three had been at the University of Sydney at the same time and knew each other. They also knew a number of people in the Department of Civil Aviation who had also been at university during the same period, and this was to prove a distinct advantage in negotiations between Qantas and the DCA.

Qantas was planning a new service to South Africa, and in 1948 QEA's Lancastrian VH-EAS made an historic survey flight across the southern Indian Ocean to Johannesburg with a group of company executives on board, including Engineering Manager Arthur Baird. It was intended that the QEA DC4 Skymasters be used to fly the South Africa service, and Bill Hill began an intensive study to determine the suitability of the aircraft for the job required. Johannesburg, which is 5000 feet above sea level, had only a small airstrip about 6500 feet long, and Bill Hill began calculating how much additional thrust was required from the DC4s Pratt and Whitney engines and what take-off performance could be expected at 5000 feet. This was in the days before calculators and computers, and all calculations had to be made using logarithm tables; under this system it could take weeks to compute one element of the mathematics involved.

Bob Walker's job was to write a Cruise Manual so that flight planning for the DC4s could be carried out. The Manual gave runway take-off lengths, landing performances and cruise performance specifications, 'so when you reached a certain altitude, you took the outside air temperature and applied both the altitude and temperature figures to a table in the manual, and this allowed you to determine the most economical cruise speed to fly, at how many revs the engines should be run, and at what pressure the manifold had to be set.' (Bob Walker)

Studies conducted by the Performance Engineers concluded that the DC4s had neither the range nor the performance for the South African service, and a decision was taken to introduce the L749 Constellations, which were known to have higher thrust and greater take-off performance, onto the route. As a result, two more Constellations (VH-EAE, named *Bert Hinkler*, and VH-EAF, named *Horace Brinsmead*) were purchased during 1950 and 1951, bringing the QEA fleet of L749s to six.

On 25 July 1952 the QEA Constellation *Charles Kingsford-Smith* left Sydney on a 'proving' flight to South Africa, flying via Perth, Cocos Island and Mauritius. Among the seventeen passengers on board were QEA's Chief Maintenance Engineer Ern Aldis, Performance Engineer Bob Walker and two ground engineers from Mascot hangars, Billy Monds and Joe Wiggins, who were to establish a new Line Station at Mauritius.

Cocos Island was of significant strategic importance to Australia's air transport communications with Europe and Africa, and recognising this the Australian Government set about providing vital wireless and weather station facilities at Cocos. Five hundred personnel from the RAAF's No. 2 Air Construction Squadron, working in conjunction with the Department of Civil Aviation, were given the task of constructing an 8000 foot runway next to the old wartime strip on the Cocos group's uninhabited West Island. The whole surface of the runway was made of gleaming hard-packed coral, which was in plentiful supply and ideal for the purpose. Huge tanks, to be used for aircraft refuelling, were installed, and a control tower built to house airport administration and a comfortable passenger terminal. QEA had a permanent staff of fourteen on Cocos Island, four of whom were ground engineers.

By May 1952 the Comet 1 had entered service on BOAC's London–Johannesburg route, and on 1 September QEA inaugurated its own fortnightly Constellation service to South Africa. Known as the 'Wallaby service' it was the world's longest commercial airline route over water, requiring the establishment of additional radio communication stations at Heard and Macquarie Islands. The flight to Johannesburg took three and a half days, and during the Perth stopover one of the Constellation's little-used Speedpak freight carriers, loaded with fresh and frozen food for Cocos Island, was attached to the underbody of the aircraft. During the refuelling stop at Cocos, the Speedpak was removed and remained on the island until it could be carried back to Perth on the aircraft's return flight from South Africa.

Meanwhile, BOAC's plans to lead the world in civil aviation were about to be dealt a series of devastating blows. On 26 October 1952 BOAC's Comet crashed on take-off at Rome, and miraculously, although it was full at the time, there were no injuries. Then on 2 May 1953 a second BOAC

Comet crashed six minutes after take-off from Calcutta, this time with the loss of all forty-six on board. A third BOAC Comet disintegrated over the Mediterranean 10 miles south of Elba while flying from Rome to London on 10 January 1954, again with no survivors. The final blow for the Comet 1 came on the morning of 9 April 1954, when a fourth BOAC Comet, with a South African crew, crashed into the sea off the Italian coast near Naples without survivors. All BOAC Comet services around the world were immediately suspended, and while exhaustive tests eventually proved that metal fatigue was the cause of the disasters, the Comet 1 would never carry passengers again.

In Queen Elizabeth's Coronation Honours List, Hudson Fysh received a Knighthood for 'services in pioneering and developing Australian Civil Aviation'. In February 1954, during the Queen's first visit to Australia, Sir Hudson was invested at an historic ceremony at Sydney's Government House.

By the end of 1953, Qantas had eight L1049 Super Constellations on order, the first of which was due for delivery in early 1954. In preparation for the arrival of the new aircraft, a number of the company's engineers were posted to the United States for specialist training at the Lockheed Technical School in airframe and engine line maintenance, hydraulics, electrics, air conditioning and flight engineering. The first eight engineers seconded to Burbank were Ron Wirth, John Unsworth, Sam Rose, John White, Alex Alexander, Don Ferme, Verne Cameron and Lloyd Hill, and they left for America during August 1953. Two members of the group, John White and Verne Cameron, also undertook an overhaul course on the TC18 DA1 turbo compound engine at the Wright Aeronautical factory in New York.

Then in January and February 1954, as the delivery date for the first QEA Super Constellation drew near, a second contingent of fourteen ground engineers and seven flight engineers was also sent to Lockheed to attend a series of practical courses on the new aircraft type. Those included in the second group were Ground Engineers Jeff Harris (from Singapore), Vaughan Sneesby (Djakarta), Bill Barrie (Manila), Noel Manning (Cocos Island), Russ Eyers and Ken Barnes (both from Darwin) and Barry Watham, Charlie Eames, Roy Carr, George Foord, Harold Avery, Ernie Fotheringham, Wal Bowry and Harold Hutton (all from Sydney). The seven flight engineers were Bruce Lawrence, Joe Silver, Ern Manning, Keith Stark, Bill Strangward, Reg Campbell and George Riek. They all undertook a four-week ground training course before starting their instruction in flight.

Although Qantas had committed itself in the short term to the American Super Constellation, it still felt the need to be kept up to date with the development and progress of other aircraft types. In August 1953, Dudley

Wright was appointed Company Representative in the United Kingdom to supervise and advise on all details connected with the Bristol Britannia project, as well as problems associated with the operation of Comet 2s on the Kangaroo route. His position as QEA's Chief Engineer in Sydney was filled by Doug Hudson.

Before the Comet 1 was finally grounded in April 1954, it was expected that BOAC would introduce their jetliners onto the England–Australia service. A reciprocal arrangement provided for QEA and BOAC personnel to service either company's aircraft on the Kangaroo route, so in December 1953, seven Qantas engineers (Col Pittard, Geoff Haddon, Alan Keats, Bruce Gates, John Cosgrove, Graham Marriott, and Kiril Bolonkin) went to the United Kingdom for nine months' training on the Comet 2. They were then posted to Singapore (Pittard and Haddon), Jakarta (Keats), Darwin (Gates) and Sydney (Cosgrove, Marriott and Bolonkin).

BOAC had already introduced the Comet 1 on its twice-weekly Hong Kong service through Singapore and Manila in March 1953. At that time George Dusting, QEA's Station Engineer in Manila, and George Foord, Sub-Foreman in Singapore, had undergone a nine-week Comet 1 handling and Ghost engine course at BOAC's Meadowbank School in the United Kingdom. George Dusting subsequently returned to England for further training on the Comets and emerged in March 1954 as the first Australian engineer to hold DCA licence endorsements for the Comet 1 and 2 aircraft and Ghost engines. He then returned to Manila to take up his position as Station Engineer. In March 1954, it was QEA's turn to reciprocate when nine BOAC engineers from that airline's line stations between Calcutta and London arrived in Sydney for a six-week training program on handling the Super Constellation, which Qantas was preparing to introduce on its Kangaroo route to England.

When Doug Hudson became QEA's Chief Engineer in January 1954, he headed a strong team of professionals, each of whom had had many years with Qantas:

Ern Aldis, Chief Maintenance Engineer and Chairman of the Apprentice Advisory Committee;
John Upex, Chief Overhaul Engineer;
Eric Kydd, Superintendent Engineering – Mascot Hangars;
Dick Ashover, Chief Technical Services Engineer;
Bill Bennett, Superintendent Engineering – Line Stations;
Jack Avery, General Engineering Superintendent.

This group of men would lead the Qantas Engineering Department into the most rapid period of expansion in the company's history, as its fleet of L1049 Super Constellations progressively built to a total of sixteen.

CHAPTER 13

A NEW SUPER STAR IN THE SOUTH

The first QEA L1049C Super Constellation (VH-EAG, named *Southern Constellation*) was delivered to Sydney 'Kingsford Smith' airport on 15 April 1954. Powered by four Wright Cyclone R-3350 TC18 DA-1 turbo compound engines, each of 3250 hp, the Super Constellation could achieve a speed of 275 miles per hour and carry up to sixty-eight passengers. By the end of May 1954 a further two L1049Cs (VH-EAH, named *Southern Sky*, and VH-EAI, named *Southern Sun*) had also arrived at Mascot.

Almost immediately, Qantas started a number of building projects. One of the first was a new engine test house, designed by a team of QEA engineers under the direction of Superintendent Repair Overhaul Shop (ROS) John Upex and Design Draughtsman James Guyot. Built on what was then the northern edge of the airfield, it was the most advanced of its type in Australia. It could accommodate both turbo-prop and the largest of the piston-type engines, and was so large (by mid-1950s standards) that it required its own fuel installation and direct water supply, the latter coming from an 8 inch bore on the site supplying 1500 gallons of water per hour for oil cooling. In the 7 July 1956 edition of *Pix* magazine, the new QEA test house was given a two-page pictorial coverage under the heading 'They Roar in Silence'. The story accompanying the photographs read:

> 'Giant aircraft engines can now be tested at Sydney Airport without noise disturbing nearby residents. The huge acoustic splinters and 14 inch-thick reinforced concrete of the new Qantas engine test house reduce to gentle hum the deafening noise of an aircraft engine running at full bore.
>
> Inside the test house, a monorail hoist lifts the engine to be tested to a dummy wing which is supported by steel pedestals bolted to massive concrete feet. The dummy wing carries control and electric cables, fuel and oil lines.
>
> Entire control, operation and observation is made by the operator [Mervyn Wass] sitting at the control panel, watching the engine through layers of sound-deadening glass. More than forty engines a month can be handled.'

While the original engine test house building is dwarfed today behind the Boeing 747 Hangars 271 and 245, it presented quite an impressive structure standing alone, as it did, on the northern side of the Sydenham–Botany goods railway line. An interesting aspect of the Test House was the fact that its internal acoustic splinter panels used for silencing cost more than the building itself.

Two nose hangars, designed to provide undercover servicing facilities for Super Constellations and Bristol Britannias, were built at Mascot during 1954 and 1955 as a temporary measure, pending construction of a major new hangar complex. Built at a cost of about £10,000 ($20,000) each, they were equipped with semi-permanent rostrums (which allowed maintenance operations to be carried out on an aircraft's nose, wings, engines and undercarriage), both natural and artificial lighting, and air, power and water services. At the time of their construction, it was intended that, when the new hangar complex was completed, the nose hangars would be demolished and sold. (One of them was relocated and stood in the vicinity of the aerodrome until 1989, when it was finally demolished to make way for the new Sheraton Hotel at Mascot.)

On 5 July 1954 Qantas took delivery of a fourth Super Constellation (VH-EAJ, named *Southern Star*). This new aircraft was an L1049'E', which differed from the standard L1049'C' in that the fuselage and wings had been reinforced by Lockheed. The strengthening was incorporated into the aircraft design to allow for possible conversion at some future date to turbo-prop engines, which would have given the Super

Constellations a 150,000 pound maximum take-off weight. (However, no civil airline ever fitted turbo-prop power plants to their Super Constellations, so the aircraft's maximum take-off weight remained at 135,400 pounds.)

Then in 1955, Lockheed further modified their Super Constellations, giving them a stronger undercarriage and a reinforced inner wing beam. All aircraft built to this standard were given the designation L1049E-01. As a result *Southern Star* (VH-EAJ) was modified to L1049E-01 standard by Qantas engineers at Mascot during 1955, and the company's original three L1049Cs (VH-EAG, VH-EAH and VH-EAI) were similarly converted during 1956.

British Commonwealth Pacific Airlines (BCPA) had been formed in 1946 by the Australian, New Zealand and British Governments to fly across the Pacific Ocean through New Zealand and San Francisco to Vancouver in Canada. By 1954 BCPA had replaced its Douglas DC4 Skymasters (five of which were sold to QEA) and was flying the larger Douglas DC6s on the Pacific route, when the Australian Government moved to acquire the United Kingdom and New Zealand shareholdings in the company. This resulted in the ultimate liquidation of BCPA and Qantas being awarded the trans-Pacific route. The first QEA service across the Pacific left Sydney on 15 May 1954. The new Qantas Super Constellation *Southern Constellation* flew the service through Nadi, Canton Island, Honolulu and San Francisco to Vancouver, inaugurating the vital Qantas–America link with an aircraft built in the United States.

Flight Engineers had first been officially listed as members of an aircraft's crew in 1947, when Qantas began its L749 Constellation services to London. By 1951 the term 'Flight Engineer' was no longer being used, having been changed to 'Engineer Officer'. Bill Strangward and Don Brown were the Engineer Officers on QEA's inaugural Pacific service, flying aboard *Southern Constellation* as far as Honolulu, where Joe Silver and Jim Forsythe took over and continued with the flight to Vancouver. QEA's L749 Constellations continued to operate the Australia–England service until August 1954, when the new Super Constellations were also introduced onto the Kangaroo route. Achievement of around-the-world services was now a possibility!

In the first three months of 1955, a further six Super Constellations (all L1049E-01 models) were delivered to QEA (VH-EAK, named *Southern Mist*, VH-EAL *Southern Breeze*, VH-EAE *Southern Moon*, VH-EAF *Southern Wind*, VH-EAA *Southern Sea* and VH-EAB *Southern Horizon*), bringing the company's fleet of L1049Es to ten. Then in March 1955 QEA bought a sixth DC4 Skymaster (VH-EBP, named *Australian Trader*) from America's Resort Airlines. It was modified to Qantas standards in the workshops of Flying Tigers and converted for use exclusively as a freighter, carrying the designation 'Qantas Air Cargo' where the name 'QANTAS' usually appeared on the upper fuselage.

The Boeing Aircraft Company of America had flown the prototype of its Boeing 707 airliner (the Dash 80) for the first time on 15 July 1954. Britain's lead in pioneering civil jet airliners had been lost. Following revelation of design deficiencies by the investigation of the Comet accidents, some of which had been carried out by Qantas engineers George Dusting and Bruce Gates, production of Comet 1s and 2s was stopped. A Comet 3 prototype — with an extended fuselage 18 feet longer than the Comet 1, a pinion fuel tank fitted to each wing and more powerful Rolls-Royce Avon engines had been built and flown successfully on 19 July 1954. It made a goodwill flight to Australia in December 1955 and created enormous interest since it was the first jet airliner ever seen in this country. However it was the only Comet 3 ever built. It was used to develop the Comet 4, which was aerodynamically identical to the Comet 3, but with a redesigned structure and new engines.

On 1 July 1955 Sir Hudson Fysh relinquished his position as Chairman and Managing Director of Qantas. C.O. Turner succeeded him as Chief Executive, and Sir Hudson continued as Chairman of the Qantas Board.

By the end of 1955, QEA had a worldwide staff of 5535. Its fleet consisted of twelve Lockheed Super Constellations (two more L1049Es, VH-EAC *Southern Wave* and VH-EAD *Southern Dawn*, had been delivered during October and November 1955), six Douglas DC4 Skymasters, eight Douglas DC3 freighters, four De Havilland DHC2 Beavers, one De Havilland DHA3 Drover and two Consolidated Vultee PB2B2d Catalinas. Work had begun on the new Head Office building in Hunter Street, Sydney, and plans were well advanced for construction of the new hangar complex at Mascot. The complex, which was to become Hangar 96, was planned as a two-bay hangar fronted by a double-storey annex and would be the largest in Australia.

Despite the success of the Super Constellations from an operations point of view, problems with the Wright Cyclone engines continued to plague the Qantas engineers. Lockheed had developed an aircraft that was beyond the capabilities of the engines powering it, particularly on long international hauls. Curtiss-Wright, the engine manufacturers, had taken the basic 2500 hp R3350-C18BD1 engine, which had powered the L749 Constellations, and increased its output to a 3250 hp power source for the L0149 Super Constellations — an increase in horsepower achieved, not by major modification to the cylinders, but by simply utilising the energy from the exhaust system. Three power-recovery turbine units were fluid-coupled to the central crankshaft and driven by the normally wasted high-pressure exhaust gases, which gave more power out through the propeller. Although this system may have seemed sound in theory, it proved to be a disaster in practice; the engine was overloaded so far beyond its capacity that it failed constantly, on occasions to the extent that its cylinders would explode out through the engine cowling. (An interesting observation was made recently by Jeff Harris, who was QEA's Acting Station Engineer at Darwin during the late 1950s. He was watching a film of a Super Constellation from Sydney coming in to land at Darwin and was moved to remark, 'Now there's an unusual sight.' Intrigued by his comment, I asked him what was so unusual about what seemed to me to be a perfectly normal landing. 'All four engines are still operating,' came his reply — and he wasn't joking!)

QEA's problem were compounded by the engine manufacturer's refusal to provide field support for their failed engines. Ron Yates recalls:

> 'Curtiss-Wright had begun losing money on their R3350-TC18-DA3, so they decided to cut their field support for the engine. Despite the extent of the problems Qantas was experiencing with its Wright engines, the manufacturer would just not give us the level of support we required and we were basically left to our own devices.
>
> Subsequently, when we were moving towards the jet age, I was asked by various aircraft manufacturers what I thought about a Wright gas turbine engine, and I had no hesitation in stating that we would have nothing further to do with Wrights. Whatever happened, we would never fly another aircraft that was powered by a Wright engine. Apparently, most airlines said the same thing, as Wrights virtually went out of the field of manufacturing engines for civil aircraft shortly afterwards.'

In the early 1950s a Production Planning Division was established within QEA's Engineering Branch. Arthur Littlejohns was appointed Production Planning Superintendent, and the stated aims of the division were to 'co-ordinate, by forward planning, the activities of the Engineering sub-departments, and to provide maximum aircraft utilisation in the most economical manner, consistent with the high degree of safety required to maintain the present reputation of QEA'.

Brian Duggan became Assistant Superintendent, and the original section leaders were Arch Graham, Cliff Roberts and Jan Aldous. The Central Planning Office was at Mascot, but the planning engineers were located at the various company workshops, from where they fed relevant information to the Central Office for co-ordination. One important part of the Production Planning process was the scheduling of aircraft movement. Don McKinven was responsible for the preparation of these schedules, which not only allowed the company to maintain its route service commitments, but also provided adequate 'ground time' for aircraft maintenance, overhaul and modification. The Scheduling Section was part of Arch Graham's Work Planning section, in which work was evaluated in terms of manhours, and estimates were compiled so that accurate schedules of aircraft ground time could be determined. The costings of the proposed work were compiled by Cliff Roberts, whose Statistics Section prepared engineering budgets and collected and classified aircraft performance information. Before work was scheduled and costed, Jan Aldous' Materials Section assessed the materials and components necessary to complete each job.

A perfect example of Qantas Engineering initiative was the introduction in early 1955 of a streamlined form of maintenance and overhaul work at Mascot. Known as the 'Block Overhaul System', the plan concentrated an aircraft's major maintenance into regular pre-determined time blocks of three weeks, and operations specification maintenance work was carried out at 2000, 4000, 6000 and 8000 hour periods. Prior to the introduction of this system, engine components were automatically overhauled at the end of their theoretical safe lifespan, as laid down by the manufacturers; this was a specific requirement of the Department of Civil Aviation in Australia. However, it was found under test that, in some cases, the amount of wear did not justify early removal and overhaul. Qantas' aeronautical engineers presented the evidence to the DCA and gained approval for an extended safe lifespan for the components involved.

The introduction of Production Planning into Qantas marked the end of the era in which judgments were made on the basis of experience. It was the beginning of a measured, scientific approach to the management of maintenance and overhaul programs, which operated in line with the concept of 'management by fact' and not by 'opinion'. As a result of thorough advance planning and rapid and methodical handling, fleet utilisation under the 'Block Overhaul System' was boosted by as much as 300 hours per aircraft per year, in effect adding one more revenue-earning plane to the QEA Super Constellation fleet.

The Super Constellations had been in operation just on two years when, in mid-1956, during a normal 2000 hour overhaul, QEA's Block Overhaul Inspector John Gundry discovered cracks in some of the aircraft's tailplane stiffening corrugations. Qantas notified Lockheed immediately, and they suggested an overall fleet inspection. X-rays revealed that most aircraft were affected, with cracks ranging in length from a quarter of an inch to three inches. Further inspection of the cracks revealed that they were localised across four of the tailplane ribs on each side, and Technical Services therefore deduced that the failure was due to metal fatigue caused by slip-stream buffeting.

A Tailplane Modification Program was introduced, which involved a general beef-up of the stiffening corrugations and the use of additional fixing brackets to reinforce the surface skin connections to the tailplane ribs. Although quite simple in principle, this process involved an extensive amount of worktime — 650 manhours per aircraft.

Flight training on Super Constellations was being carried out at Narromine airport in New South Wales. It was becoming an increasingly costly operation, so Qantas purchased and installed a Redifon Flight Semi-Simulator at the Technical Training Centre at Mascot. Although somewhat primitive compared to the Boeing 747 Flight Simulators of today, the Redifon was able to simulate actual airborn conditions without leaving the ground. It could seat two pilots in an exact replica of an L1049 cockpit, while an instructor sat behind the pilots working the control panel, simulating a number of actual inflight situations. Pilots were able to taxi, take off, manoeuvre and land while hearing the noise of the engines, wind and touch-down tyre squeal. They could also see instruments respond to altitude and temperature change; feel the controls react to airspeed attitude of the aircraft and position of wheels and flaps (retracted or extended); or crash the trainer if they failed to manoeuvre correctly. Some procedures still required practical application in an actual aircraft (for example, landing with one engine shut down), but the cost of flight training was greatly reduced by the Redifon.

In 1956 QEA became the first airline to order the L1049'H' model of the Super Constellation. It was a convertible aircraft, designed to be switched quickly from passenger to cargo use as traffic required. QEA purchased two L1049Hs (VH-EAM, named Southern Spray, and VH-EAN, named Southern Tide) and placed them into service on the Pacific route in November to help handle passengers travelling to the Melbourne Olympic Games.

QEA's involvement in the 1956 Olympic Games was not restricted to the carriage of overseas passengers to Melbourne. The Olympic Flame had to be carried from Athens to Australia, and the QEA Drawing Office at Mascot was given the task of designing a suitable container for carrying the Flame in one of the company's Super Constellations. The problem was, of course, how to carry the Flame aboard the aircraft within the limits of safety and keep it alight on such a journey. After much experimentation, Jim Allan and Phil Ballard designed a special 'Olympic Torch Unit', comprising an asbestos-lined box in which a Saar miner's lamp was suspended and protected against buffeting by a series of springs; a rack and stand for the Ceremonial Torch and a built-in fire extinguisher. The Unit was mounted into a frame that normally held one of QEA's Skycot baby's bassinets and was attached to the front wall of the forward passenger cabin of the Super Constellation. Initial tests on the miner's lamp to find out how long it would burn were carried out using kerosene. But it was finally decided that methylated spirits was the most suitable fuel, because it would keep the lamp alight for thirty hours without needing to be refuelled between the aircraft's various ports of call along the route to Australia.

The Torch was ceremoniously lit from the rays of the sun on the Plain of Olympia and carried by relay runners to the Acropolis in Athens, where it was given into the care of QEA's Captain Andy Young and Office Controller Col Porter, whose function was to guard the flame on its 8600 mile flight. Upon arrival at the QEA aircraft, the flame was transferred from the Ceremonial Torch to the miner's lamp, which was then fixed inside the asbestos-lined container. The Torch was now extinguished and placed in position on the special rack and stand in the unit designed for its carriage.

The flight to Darwin, via Calcutta, Bangkok, Singapore and Djakarta, was achieved without incident in the elapsed time of 52 hours 20 minutes (actual flying time Athens–Darwin 31 hours 10 minutes). To identify the aircraft as the Torch carrier, Qantas engineers had also designed and manufactured a small flagpole, which, with the Olympic Flag, was attached to the port side of the aircraft when it taxied in and out of each stopping-off point along the route. At Darwin the Olympic Torch Unit was transferred to an RAAF Canberra bomber and flown to Cairns, where the Torch was reignited and handed over to relay runners for its journey down the east coast of Australia to Melbourne.

At approximately 8 am on 17 November 1956 (Sydney time), QEA's Super Constellation VH-EAH was on a routine flight between Nadi and Honolulu when a cylinder failure in No. 2 engine caused it to be diverted to Canton Island. It was a Saturday morning, and at 10:10 am Mascot Ground Engineer Alan Woods was sitting in a barber's chair at Kogarah Bay having his hair cut. Before an hour had elapsed, he was airborne on his way to Canton Island, together with Ted Wickham (Shift Foreman Constellation Servicing, Hangar 85) and Ken Pearson (Constellation Servicing). These men were members of Bill Bennett's 'Flying Squad', a group of twenty volunteer engineers who could be called upon by the Superintendent Route Stations – Engineering (Bill Bennett's title at the time) to go anywhere, anytime, and at a moment's notice, when the company was faced with an emergency situation.

By 12:30 pm the Qantas DC4 VH-EBK had been withdrawn from the New Guinea service, and it too was in the air headed for Canton Island with a replacement R3350-TC18-DA3 engine and two additional engineers, Cec Knott and Les Peates (also members of the 'Flying Squad') on board. The DC4 flew the 3291 statute miles to the small Pacific atoll in 15 hours 30 minutes and, with three hours' fuel reserve, arrived at Canton Island at 7 am on Sunday morning, three hours behind Ted Wickham and his group. This was the longest flight undertaken by a Qantas DC4.

Ted Wickham, an ex-Qantas apprentice with fifteen years' service with the company, commanded the actual engine change-over. His report at the time stated in part:

> 'On arrival at 4:00 am on Sunday morning, I found nothing had been done to the aircraft. The only crane available was some distance from the aircraft and could not be moved. Both the DC4 and the unserviceable 1049 had to be taxied up to and taken away from the crane in turn, so that the engines could be lifted in and out.'

Despite this difficulty, and the fact that working conditions were complicated by extreme heat and the glare reflected from the white coral surrounding the airstrip, QEA engineers had the engine changed and running on the delayed Super Constellation by 6 pm, just eleven hours after the arrival of the DC4. It is reported that when Ted Wickham was later asked to comment on the operation, he simply shrugged his shoulders and said, 'It was just another engine change!'

By the end of 1956 work on the new Hangar 96 complex at Mascot was well advanced and nearing completion.Then on 12 January 1957, Bay 1 collapsed, suddenly and without warning. Fortunately no-one was injured and no aircraft were in the hangar at the time, but the collapse caused significant delays to the company's program of planned maintenance and major overhauls of Super Constellations in the hangar complex.

Early in 1957, the new L1049H *Southern Spray* became the first QEA Super Constellation to have weather-mapping radar equipment installed in its nose. The radar was able to 'feel out' 150 miles ahead of the aircraft, detecting storm centres and aerial turbulence. Installed by QEA engineers and AWA radio technicians at Mascot, the equipment consisted of a transmitter–receiver unit, an accessory unit, a control panel and an antenna and indicator. The antenna was located in the aircraft nose behind a fibreglass 'radome'. The type chosen was the RCA AVQ 10 'C' Band system, which was already being used extensively by American and European airlines. It was a commercial development based on experience gained with military terrain-mapping radar. American military planes had been using the equipment for several years to avoid dangerous turbulence in bad weather. Its simple operation was described as follows:

> 'A pencil beam is transmitted from the rotating antenna in the nose, and its energy is partially reflected back to the antenna by either moisture-laden clouds or rainfall. The amount of energy reflected is proportional to the density of the moisture in the target, and is displayed on a screen in the flight station. In this way, the pilot can 'see' clouds one behind the other, even through intervening rain, and can choose to fly over or around the turbulence.'

By the end of 1957, the entire QEA Super Constellation fleet had been equipped with weather-mapping radar.

Another important modification to the Qantas Super Constellations was the fitting of wing tip fuel tanks to the final two of the aircraft type purchased by the company during 1957. These aircraft were both L1049Gs (VH-EAO, named *Southern Aurora*, and VH-EAP, named *Southern Zephyr*). The tip tanks provided an additional 1000 imperial gallons fuel capacity and a further 650 miles range, enabling the Super Constellations to fly the 3215 miles between Fiji and Honolulu without the necessity of a refuelling stop at Canton Island. The extended range also enabled them to fly the Atlantic Ocean from New York to London. QEA engineers at Mascot subsequently modified five existing L1049Es (VH-EAA, VH-EAB, VH-EAC, VH-EAD and VH-EAE) to L1049G standard, using special kits purchased from Lockheed.

On 28 October 1957, Prime Minister Menzies officially opened the new company headquarters. the impressive Qantas House building in Hunter Street, Sydney. By this time, Qantas had grown to become the seventh largest international airline in the world.

In the New Year's Honours List in 1958, QEA's Chief Executive and General Manager, C.O. Turner, was honoured with an OBE in recognition of his twenty-three years' service to aviation in Australia. Then on 14 January, Qantas inaugurated its long-awaited round-the-world service from Essendon airport in Melbourne with its two new Super Constellations, *Southern Aurora* and *Southern Zephyr*. The flight was to make aviation history. Staff gathered on the roof of the new Qantas House in Sydney as the two aircraft flew past in formation before breaking away to circle the world in opposite directions, *Southern Aurora* flying eastward and *Southern Zephyr* westward. It was the beginning of two complete round-the-world services each week for QEA, one from Melbourne and the other from Sydney.

In early March 1957, BOAC commenced operations with their Bristol Britannias on the London–Sydney route. In anticipation of this, a number of Qantas engineers went to England during 1956 for extensive training on the Britannias. Engineers known to have undergone Britannia training in England were George Dusting, Bruce Gates and Darcy Dumbell (all from Mascot), Jeff Harris and Russ Eyers (from Darwin), Keith Pearce and Stan Quigg (from Djakarta), Graham Marriott and Geoff Haddon (from Singapore) and Alan Keats (from Manila).

Qantas had determined by this stage that it would not buy Bristol Britannias for its own services but would continue to operate Super Constellations until the introduction of jet airline services, which were scheduled to commence in 1959. This proved to be a good decision for Qantas, since before long the Britannias began to suffer a spate of problems with their Proteus turbo-prop engines. To reduce the length of the power plant, the Proteus used a reverse flow intake system; air entered an annular intake behind the propeller spinner and was drawn along ducts to a point halfway between the propeller and the wing, where it turned through 180 degrees and flowed forward to the compressor. Ice was found to build up at the bend in the intake and then break away in lumps, which would damage the compressor blades or cause the engine to flame out. It would be mid-1958 before extensive testing and modifications to the engines would solve the problem.

By early 1959 an enlarged and redesigned Comet aircraft had at last begun successful operation, and BOAC ordered a substantial fleet of the new Comet 4s. But success came too late for the Comets. Qantas was now confident in its ability to select aircraft for its fleet with complete independence and on the strictest criteria of technical and commercial merit. It had chosen Boeing 707 jet aircraft to replace the Super Constellations on its international routes, and in September 1956 had placed an order for seven Boeing 707-138s for delivery during 1959.

Thus began an era of technical expertise never before experienced in Australian aviation, where the Qantas engineers would lead the world in acheiving standards of excellence and integrity of workmanship. This was the era of the Qantas Jet Base.

CHAPTER

14

THE BOEING DECISION: A MANDATE FOR EXPANSION

When the first Qantas 707 was towed out of Boeing's Renton plant in Seattle on 16 February 1959, it was the first American-built jet airliner to be delivered to a non-American airline. The 707-138 (VH-EBA, named *City of Canberra* at the time of roll-out) was the Qantas version of the standard Boeing 707-120, but was faster and had improved all-round performance, particularly on take-off. A water injection system in its four Pratt and Whitney JT3C-6 turbo-jet engines used 2500 to 5000 pounds of water on take-off, depending on conditions, and added about 7 per cent take-off power.

By the end of 1959, QEA had a fleet of seven Boeing 707-138 jet aircraft operating on its round-the-world services: VH-EBA, renamed *City of Melbourne* in July 1959; VH-EBB, named *City of Canberra* for delivery, then renamed *City of Sydney* in July 1959; VH-EBC, named *City of Melbourne* for delivery but renamed *City of Canberra* in July 1959; VH-EBD, *City of Brisbane*; VH-EBE, *City of Perth*; VH-EBF, *City of Adelaide*; and VH-EBG, *City of Hobart*). It was actually VH-EBC, carrying its new name *City of Canberra*, that made the inaugural jet flight across the Pacific from Sydney to San Francisco on 29 July 1959. The despatch of the aircraft from Sydney's Kingsford-Smith Airport was a colourful ceremony, attended by the Minister for Civil Aviation, The Hon. Shane Paltridge and the Minister for Air, The Hon. F.M. Osborne.

By the time Qantas placed orders for its seven Boeing 707s in September 1956, the company had explored several options for re-equipping its fleet. Initially it was thought that it would be 1960–61 before Qantas could enter the jet age with aircraft suited to its operations, and at that time the company was basing its forward plans on the Douglas DC8 jet airliner. But to hold its position on world routes between 1957 and 1961, Qantas had to explore the possibility of introducing an interim aircraft type into service. Deliberations revolved around the Douglas DC7-C, the Lockheed 1649A and the Bristol Britannia. (The Comet 4 was clearly being regarded as a short- to medium-range aircraft at this stage of its development and therefore unsuitable for Qantas route requirements. Besides, delivery of the Comet 4 could not be guaranteed before 1959 at the earliest.)

Qantas saw the Bristol Britannia as an aircraft designed for operation in the 1953–59 competitive era, but by 1956 the aircraft had still not entered regular service and Qantas did not feel it could compete successfully in the late-1950s with a 1953 model. Interestingly, BOAC had to purchase ten Douglas DC7-Cs for its New York–London service to tide them over as a result of continuing delays with the Britannias. In March 1957, BOAC finally introduced Britannias on the Kangaroo route, seven months after Qantas had committed itself to Boeing 707s.

The Douglas DC7-C was an advanced version of the standard Douglas DC7, developed at the request of Pan American Airways, which wanted it for their Pacific Ocean service to Australia. Known as the 'Seven Seas', the DC7-Cs featured a greater operating range of just over 5000 miles and cruising speed of 355 miles per hour. They represented a potential threat to Qantas operations on its Southern Cross route across the Pacific, because they flew 25 miles per hour faster than QEA's Super Constellations. From a commercial point of view, this would have made Qantas uncompetitive with Pan Am, so consideration was given to the Douglas DC7-C as a possible interim aircraft type for Qantas operations.

Then Lockheed announced the production of its 1649A, which had been developed for Trans-World Airlines (TWA) in direct competition with the Douglas DC7-C. According to Lockheed publicity, the 1649A was 'the longest range aeroplane in the world, capable of non-stop flights up to 6300 miles without using its fuel reserves. Its 150 foot wing carries 9600 gallons of fuel and allows relocation of the engines farther outboard for least cabin noise level. The 1649A wing is one-sixth thinner [than the 1049H] making possible cruise speeds of 350 miles per hour and a top speed of nearly 400 miles per hour.'

But Qantas soon found that it would not be able to put either the Douglas DC7-C or the Lockheed 1649A into route service in effective numbers before mid-1958. Ron Yates recalls that, despite these late delivery dates,

> 'There was a great deal of enthusiasm within Qantas to go with the 1649, since we already knew Lockheed very well and commercially we enjoyed a very good management relationship with them.
>
> But Scotty Allan (who was then QEA's Assistant General Manager) and I had a completely different point of view. We had come to be very impressed with Boeing's knowledge of jet aircraft operation, as well as their background and experience. We also knew that Douglas did not have any depth of experience in jets at that point of time, and production of their DC8 was running about a year behind Boeing's 707 project.
>
> The Boeing Company had had a lot of experience with jets because of their military involvement. They had people involved with the occupation forces that went into Germany after World War II, and they took the opportunity to study German research. As a result, they concluded that the sweep back on the wings of jet aircraft ought to be at thirty-five degrees, and Boeing ultimately changed its wing design to achieve a thirty-five degree sweep back.
>
> We were faced with the option of going with the Boeing 707 in 1959 or waiting until 1960–61 for the Douglas DC8 and introducing either the Lockheed 1649A or the Douglas DC7-C as a stop-gap measure.
>
> So Scotty and I said, 'To hell with that! Why don't we push to make the 707 the type of aircraft we need. If we buy that aircraft, we will not have to adjust to the operation of two new aircraft types within a relatively short period, and we can scoop the pool by getting into the jet age a year earlier.'

The decision to go with the Boeing 707 was taken, and Government approval for the purchase was announced by the Minister for Civil Aviation, Athol Townley, to the press gallery at Parliament House in Canberra on 6 September 1956. But there were many months of negotiation with Boeing still to be undertaken before Qantas would take delivery of its first 707. Ron Yates again:

> 'Boeing was under great pressure from a number of airlines, including Pan Am and TWA, to stretch the 707 because they wanted it slightly bigger. So Boeing decided to put an extra ten feet into the length of the aircraft and in doing so, absolutely

> destroyed its payload capability from our point of view on the Pacific, and that was where we wanted to use this aircraft. After much negotiation, Scotty and I finally reached agreement with Boeing to take ten feet out of the Qantas aircraft, so our 707 was built to the original length of 128 feet 10 inches instead of 138 feet 10 inches. This version was designated the 707-138 model and became widely referred to as the "Qantas Boeing" '.

It was also discovered that the Pratt and Whitney JT3C-6 engines did not have sufficient thrust for take-off in high temperature conditions at Fiji, where the runway was only 7000 feet long. Boeing improved the leading edge devices on the Qantas 707s to provide more lift for take-off at Nadi, but that alone did not solve the problem. Ron Yates:

> 'We needed additional thrust up to 85 degrees fahrenheit when the thrust normally fell off quickly after 75 degrees, so after a lot of negotiation with Pratt and Whitney, they agreed to permit Qantas to use the thrust at the same level right out to 85 degrees before reducing it, just for take-offs at Nadi. It produced what became known as "The Nadi Bump", and that was a bump in the thrust curve which only occurred during take-off in Fiji.'

In October 1956 Ron Yates (who was Qantas Projects Engineer at the time) moved to Seattle to establish a company engineering representation at Boeing's Renton plant. He returned to Australia in early 1957 and became Project Controller for the Boeing 707, a position that gave him total command of the project. He was responsible only to senior Qantas management, was given and complete leeway to handle all aspects of the project, including commercial, operations and engineering; his co-ordination of the entire undertaking exposed him to every area of the company's operation. John Mills, a former Chief Inspector at the Mascot Hangars, replaced Ron Yates as Resident Engineer at Boeing in February 1957.

Preparations for the arrival of the jets had had begun at Mascot in 1956. As soon as Qantas had made known its intention to purchase the Boeing 707, Chief Executive C.O. Turner announced a major program of construction at what was to become known as the Qantas Jet Base. The £6 million ($12 million) scheme included a new Engineering Repair and Overhaul facility, a jet engine Test House, a 707 Line Maintenance Hangar with two-storey workshop annex, and a Technical Training Centre.

Thus began a period of unprecedented challenge for the Qantas engineers. The first step in positive engineering planning was the formation of special working groups in December 1956. The Senior engineers who headed the groups focused on the Boeing 707 aircraft, the Pratt and Whitney JT3C-6 engine, appropriate methods of staff training, and the selection of suitable ground-handling, maintenance and engine overhaul support equipment for the introduction and operation of revolutionary new aircraft type. It took nine months for the groups to crystallise their proposals, and in September 1957 an Engineering Production Co-ordination Committee was formed under the chairmanship of Ern Aldis (Assistant Chief Engineer). Other members of this committee were Eric Kydd (Chief Maintenance Engineer), Bob Scott (Materials Superintendent Bulk Store), John Upex (Chief Overhaul Engineer), and Jack Avery (General Engineering Superintendent). Bernie Bentley (Technical Publications Officer) was the Committee Secretary. The Co-ordination Committee's function was either to ratify the decisions and recommendations of the working groups or refer contentious points back to them for further study. The Committee met once a fortnight, but could be called together at short notice if an important issue had to be discussed.

A Boeing Production Group, chaired by Brian Duggan (Assistant Production Planning Superintendent), implemented the directions issued by the Co-ordination Committee and controlled the manufacture or procurement of items needed for the 707 project. Much of the equipment and tooling was designed and drawings prepared by the Drawing Office in the Qantas Technical Services Department at Mascot, under the direction of Chief Draughtsman Jim Allan, and then manufactured either in Qantas' own workshops at Randwick or by outside organisations. The specialised tooling equipment for the machine shop was also manufactured in Australia but from blueprints supplied by Pratt and Whitney.

In mid-1957 development began on the floor plan for the new Repair and Overhaul Shop (ROS) at Mascot, which had to provide 'overhaul support for complete power plants, all instrumentation, electronics, communication equipment and radio aids, electrical and mechanical components'. By the end of August the floor plan details were complete, and architects had the structural design ready for the builder to start work in November. It was built on a site north of the Sydenham–Botany goods railway line, immediately west of the Qantas Passenger Services building, and covered an area of 140,000 square feet.

In the preceding years, Qantas had developed a highly specialised overhaul system to meet the requirements of the Super Constellation's piston engines, but it now became apparent that jet engines would demand an appreciably different type of specialisation, and that support equipment held for the piston engines would have little application in the gas turbine overhaul field. Since Qantas had no plans to bring its piston engine program to an abrupt end when its jets began operating in 1959, an overhaul system suitable for both types of engines had to be developed for the new ROS, a mammoth task overseen by Chief Overhaul Engineer John Upex. A workshop layout was developed so that both gas turbine and piston type engines could be overhauled through the same flowlines. In an article published in the October 1961 edition of *Aircraft* magazine, John Upex wrote:

> 'This required detailed study to ensure that both the fixed and handling equipment, such as dis-assembly, rework and sub-assembly stands could, by adaption, be made to take the gas turbine engines. Wherever possible, we adapted piston engine stands to take the gas turbine engines. Where this was not possible, we modified the gas turbine stands to cover the requirements of the piston engines. However, our studies clearly established that in the final assembly stage, equipment could not be conveniently developed to handle both piston and gas turbine engines along a common assembly line.'

The new facility was ready for occupation by October 1958, but since it was necessary to maintain complete overhaul support for the company's existing fleet of sixteen L1049 Super Constellations, the transfer of various workshops was spread over six months.

Assessment and selection of suitable overhaul and functional test equipment for the Boeing 707s made many demands on the ingenuity of Qantas engineers. Being the first airline to base American-built jet aircraft outside the United States, Qantas' component and overhaul support requirements were (by necessity) well ahead of delivery dates being offered by the manufacturers of specialised overhaul support equipment. A further complication was the fact that it was common for manufacturing organisations to develop highly specialised functional test equipment to suit their own production requirements, and since overhaul processes usually followed a pattern different from that of manufacture, the manufacturing support equipment frequently did not fill the needs of overhaul. This added considerably to the Qantas engineers' work in selecting equipment, and to meet deadlines they were often left with no alternative than to design, develop and manufacture their own support equipment.

The most complex piece of test equipment developed and manufactured by Qantas engineers was a functional test bench for the Bendix PB20D Automatic Pilot system fitted to the Boeing 707s. The test bench was installed in a special soundproof area, isolated from the high level of noise emanating from the workshop, which was established specifically for testing electrical components and navigational equipment.

After many months of extensive research and careful consideration, during which lengthy discussions were held with Pratt and Whitney, decisions were finally taken on equipment purchases for the new ROS. Schenk Dynamic and Trebel Rotary Static Balancing machines were purchased from Germany, while the machines for moment mass weighing of individual engine blades came from England. Two Nankervis gas turbine fuel control and fuel injection test benches, which combined the fuel booster pumps, fuel flow transmission and fuel flow indicators into the functions of a single bench, were also purchased and installed at a cost of approximately £50,000 ($100,000) each. The rates of flow demanded flow meters to cover a range of up to 20,000 pounds per hour, and the accuracy required of flow measurement left little alternative but to use electronic digital counters for presenting rates of fuel flow.

Equipment for testing the more dynamic components was installed in special cells in an annex attached to the main workshop. The annex consisted of two main rooms, each with a floor area of 1600 square feet, and six independent cells. In one of the main rooms, all the heavy equipment that did not require special fire-protection considerations was installed, including hydraulic pumps and motors, water boost pumps, water injection pumps and speed governors. The other main room contained components categorised as requiring fire protection, and it was in this area that the two gas turbine fuel control system calibration benches were located.

During construction of the new ROS, Qantas announced its intention to purchase four Lockheed L188C Electras, to be brought into service in late 1959 on the QEA secondary routes to Hong Kong and Tokyo, and later to New Zealand, Noumea and Fiji. The Electras were powered by four General Motors Allison 501-D13 turbo-prop engines of 3750 hp each. They were actually a derivative of the T56A-11 military engine on the Lockheed C130A Hercules transport aircraft, and with the Government approval for Qantas to purchase its Electras came a decision in favour of the C130s for the RAAF. Ansett-ANA also wanted to order four Electras for its domestic services in Australia, but was restricted by the Commonwealth Government to only two. In contrast, Australia's other domestic carrier, TAA, wanted no part of the Electra but was virtually forced into purchasing two at the insistence of the Government.

In October 1958, three QEA engineers from ROS Mascot — Don Mealing, George Ball and Colin Clark — went to the Allison engine plant at Indianapolis for training on the new engine type. Don Mealing and George Ball also went to Bendix at South Bend, Indiana, to study the fuel system on the Allison engines.

When John Upex pulled the main power switch to 'on' at the new ROS on 7 October 1958, it was planned that by 1960 the facility would be overhauling 330 gas turbine engines per year: 120 Pratt and Whitney JT3C-6s from the Boeing 707s, 50 Allison T56-11s from the RAAF's Hercules, and 160 Allison 501-D13s from the Electras (including those belonging to Ansett-ANA and TAA), and of course the Wright Turbo

Compound R3350-TC18-DA3s from the nine remaining Qantas Super Constellations.

In early 1959 Pratt and Whitney provided Qantas with a 'bare' JT3C-6 engine, so that a pilot overhaul run could be made through the new ROS, and a test run could be undertaken in the recently completed jet engine Test Cell. The Test Cell was built on the north-western boundary of what was then being referred to as the Qantas Industrial Area at Mascot. It was one of the most advanced jet test cell designs in the world, and overall control of the project fell under the jurisdiction of Chief Technical Services Engineer Dick Ashover. Power Plants Engineer Bob Bennett did the actual detailed design work, and General Engineering Superintendent Jack Avery supervised the construction, acting as co-ordinator between Qantas design engineers and the building contractors. The layout of the new installation was planned to include:

(a) The Test Cell itself, constructed in reinforced concrete with walls 16 inches thick and a roof thickness of approximately 12 inches. The cell had a single vertical stack air intake, fitted with specially designed acoustic splinters. It could accommodate turbo jet engines of up to 30,000 pounds static thrust with afterburning and was designed to include a Durastack silencing system in the exhaust assembly to bring the generated noise down to an acceptable level. (b) A control room with an observation window and acoustically treated doors, from which all functions of the engine under test could be monitored. (c) An equipment room which contained the fuel plumbing and fuel measuring systems, as well as the compressed air and electrical systems (Both the control room and equipment room were constructed in cavity brickwork, with reinforced concrete floors and ceilings. No physical contact was made with any other part of the test house structure except through isolation joints, thus preventing the transmission of sound and vibration from the test cell. (d) An engine preparation shop, where the engine was 'dressed' before testing. It was a steel-framed area lined with coloured asbestos cement and glass.

The Fuel Tank Farm, which was built in conjunction with the Test Cell, contained three tanks of 12,000 gallons each. When the Test Cell building and its mechanical installations had reached an advanced stage, Sub-foreman Jan Aldous and a team of engineers and electricians moved in to install the controls and instrumentation. The electrical wiring from the console to the test engine was designed by QEA Electrical Engineer John Cotton.

Although Qantas engineers were prepared for problems, they encountered no hitches in the initial fitting of the special test engine. 'The engine fitted like a glove,' was reportedly the pleased comment from Chief Engineer Doug Hudson. The fault-free run of the engine in the new Jet Engine Test Cell proved that, after many years of intensive planning, Qantas engineers were now ready to enter the Jet Age.

The original piston-engine Test Cell at Mascot was adapted to handle the Allison engines from the Electras and Hercules by the same design and installation team responsible for the jet-engine Test Cell. The first turbo-prop power plant was successfully tested in the converted facility during March 1959.

To facilitate the efficient management of Boeing 707s at Mascot and at certain of the company's line stations, new ground handling equipment had to be designed and manufactured. Overall responsibility for this rested on General Engineering Superintendent Jack Avery. Seven special models of the British Douglas Tugmaster tractors were bought at a cost of £6888 ($13,776) each. They had been built to specifications developed jointly by Qantas and Douglas Equipment Limited of Cheltenham, England, to meet the precise requirements of Qantas. Fitted with a 230 hp six-cylinder Leyland diesel engine, they were capable of towing or pushing aircraft to a maximum all-up weight of 300,000 pounds. A 30 KVA petrol-motor-powered generator unit was mounted on the right-hand side of the Tug to provide power required on the 707s during pre-flight handling. One of the tractors was retained for use in Sydney while the other six were sent to Avalon (Victoria), Rome, Singapore, Nadi, Honolulu and San Francisco.

Thirty-three mobile stair units, built by Hastings Deering at Lidcombe in Sydney to specifications drafted by Jim Allan of Qantas Technical Services, were delivered ready for use prior to the arrival of the first Qantas 707. They were built in sections so they could be easily dismantled and transported to Line Stations. George Roberts (who had moved from the field of instrumentation) was the Engineering Equipment Officer who oversaw the manufacture of the mobile stairs:

> 'The stairs had to be designed to carry quite a load, since we found that football teams and athletes travelling overseas to Olympic and Commonwealth Games invariably posed for the cameras on the stairs with the aircraft in the background. So we allowed for each stair to carry the weight of two people simultaneously.
>
> The design also allowed for the stairs to be dismantled for transport to our other bases throughout the world, and then reassembled in a period of eight hours — one set of stairs could be carried on a DC3 and two sets on a DC4.
>
> We used Ford parts in the manufacture of the mobile stairs for the very good reason that, wherever they may be located in the world, spares were usually available as automotive parts off-the-shelf.'

Although designed primarily for operation with the Boeing 707s, the mobile stairs could be raised or lowered for use with any of the aircraft types in the Qantas fleet. When they were driven up to an aircraft, hydraulic rams were lowered on each of the four corners, which lifted the wheels off the ground. This not only gave the mobile stairs stability, but prevented the possibility of the tyres puncturing and causing the unit to lurch to one side, inflicting damage to the aircraft.

On 13 April 1959 the Minister for Civil Aviation, Senator Shane Paltridge, officially commissioned the new Qantas million-dollar Link Jet Flight Simulator at the Mascot Technical Training Centre. It was technically an exact replica of the Boeing 707 cockpit and gave a faithful reproduction of all flying conditions, simple or complex. While engineers Brian Murton, Harry Stringer and Pat Ovens kept a watchful eye on the operation of the simulator equipment, guests invited to the launching saw a Qantas Flight Crew (which included Engineer Officer Arthur Rowe) simulate a jet flight from San Francisco to New York. Such was the interest in the event that it was featured as an item in the Fox Movietone newsreel shown in cinemas throughout Australia the following week. (Five years later Qantas became the first airline in the world to install a 'digital' flight simulator with a built-in motion system able to produce all the sensations of *movement* in flight not previously possible with the older-style 'analog' simulators.)

In March 1959, the Qantas Board approved a new Company Organisation Chart, which incorporated a number of key changes in the senior executive positions. Captain R.J. (Bert) Ritchie became Director of Technical Services and Doug Hudson was appointed Engineering Manager, responsible to the Director of Technical Services for the direction of engineering functions, including technical development and services.

The Engineering Branch was divided into two main departments: Engineering Maintenance and Overhaul, and Technical Development Department. In the Engineering Maintenance and Overhaul Department, Ern Aldis was made Chief Engineer and Dick Ashover Assistant Chief Engineer. According to the special edition of *Qantas News* dated March 1959, which announced the details of the company reorganisation,

> 'The build-up of our engineering production flow in aircraft overhaul, engine overhaul and testing, component electrical and electronic overhaul — allied with the introduction of new aircraft types — has called for a strengthening in this area.'

Accordingly, the departments were divided into three main divisions: Engineering Production, Aeronautical and Services Engineering, and Inspection. The Engineering Production Division was headed by John Mills, who took up the position of Engineering Production Manager in August 1959, after returning from two and a half years in the United

States as the company's Resident Engineer in Seattle. Eric Kydd (Chief Maintenance Engineer), John Upex (Chief Overhaul Engineer) and Arthur Littlejohns (Chief Planning Engineer) all became directly responsible to the Engineering Production Manager.

Bill Hill became Chief Aeronautical and Services Engineer, responsible for the provision of specialist technical services to the Maintenance and Overhaul Division, and Con Gehrig retained his position as Chief Inspector.

In the Technical Development Department, both Dr R.R. (Dick) Shaw (the Technical Development Controller) and Ron Yates (the Boeing Project Controller), reported directly to Engineering Manager Doug Hudson. Bob Walker (the Electra Project Controller) was responsible to Dr Shaw.

On 5 September 1959, Qantas 707s began flying from Sydney to London by the Southern Cross route to San Francisco, then on via New York and across the Atlantic. Then on 27 October, VH-EBE *City of Perth* inaugurated the first jet air service between Sydney and London along the Kangaroo route. The flight took 37 hours 15 minutes and linked with the Qantas service through the United States, thus providing the first round-the-world 'jet' service through New York.

Qantas took delivery of its four Lockheed L188C Electras during November and December 1959: VH-ECA, named *Pacific Electra*, VH-ECB *Pacific Explorer*, VH-ECC *Pacific Endeavour*, and VH-ECD *Pacific Enterprise*. They replaced the Super Constellations on the Qantas Far East routes on 18 December and reduced the flying time from Sydney to Tokyo (via Darwin, Manila and Hong Kong) by 5 hours 45 minutes. Super Constellations continued to operate passenger services to South Africa, but from the beginning of 1960, flew to the Far East only as cargo carriers.

CHAPTER

15

OUT ALONG THE LINE

Line Stations have always been an essential ingredient in the overall success of Qantas operations. In the beginning they were simply petrol and oil depots set up along the way at various towns in outback Queensland where the aircraft landed, to provide refuelling stops for the fledgling airline. (Note: At various times in the company's history, Line Stations were also known as 'Out Stations' or 'Route Stations', but in the interests of consistency only the term 'Line Station' is used throughout this chapter except where specific titles are referred to.)

When Qantas Empire Airways began its overseas operation from Brisbane to Singapore in January 1935 with Commonwealth-class DH86s, additional manned Engineering Line Stations were established along the route at intervals within the range of the aircraft at Darwin,

Koepang, Rambang, Sourabaya, Batavia and Singapore. The end of World War II saw aircraft with far greater range introduced to the Qantas fleet, and coupled with the rapid expansion of its air routes this resulted in a change in the company's requirements for its Line Station locations.

Dave Thomson had become Qantas Area Engineer at Darwin in 1945, after having served at the company's Line Stations at Sourabaya and Townsville with Eric Kydd during the war. Qantas was about to introduce three Hythe-class flying boats and three Lancastrian land-plane services per week to London in association with BOAC, and Darwin was a night stopover on both services. The company took over the wartime 117th Army General Hospital at Berrimah, 9 miles south of Darwin, and converted it into a village to provide accommodation for airline staff and passengers. The fully self-contained village had a staff mess, passenger lounge and hospital facilities, as well as a half-Olympic-size swimming pool, which offered welcome relief from the heat for many a weary passenger and Qantas employee.

The November 1947 edition of *Empire Airways* magazine described the new Qantas facility at Darwin for the benefit of airline passengers:

> 'Catering for fifty-three passengers, Berrimah is laid out on holiday camp lines. Surrounding bushland contrasts with neatly arranged gardens in which tropical fruits such as bananas and pawpaws are prominent. Many of the rooms are equipped with their own showers, and construction throughout is directed especially at achieving coolness in the hottest weather. Passengers have a large reading lounge, lounge bar, and a dining room with high-quality food.
>
> In course of erection is a large aviary to be filled with Australian birds (always of interest to visitors from overseas). In addition to passengers, aircrews and staff are accommodated at Berrimah, present staff totalling about 120.'

A new hangar complex was constructed by the Department of Civil Aviation at Darwin airport to replace the original Qantas hangar that had been destroyed during enemy air raids in February 1942. The new hangar provided engineering facilities at one end and a passenger lounge at the other. The engineering staff, including the female stenographers who worked for Dave Thomson, fitted out the new hangar offices and workshops to suit the Qantas operation. In 1948 they also designed and built a garage and workshop at Berrimah to house the company's fleet of buses and cars.

Since Darwin was a night stopover Station, the engineers had from 8 pm to 6 am to service the Lancastrians and Hythe Flying Boats and rectify any problems reported by air crews before the scheduled early-morning departure time. But from December 1947, when the L749 Constellations commenced operation on the Kangaroo route, Darwin ceased to be a night stopover Station. The Constellations would leave Sydney at 9 pm and fly across Australia through the night, arriving at Darwin in the early hours of the following morning for a quick transit stop, which left the engineers just two hours to service the aircraft and carry out any necessary repairs. Passengers were transported by Qantas buses to Berrimah, where they were given breakfast and the opportunity to shower and shave before being returned to the aircraft for a direct daytime flight to Singapore. The service was scheduled to arrive at Singapore at 3:30 pm, where it would remain overnight before proceeding on to Karachi at 7 am the next day.

With the introduction of the Hythe Flying Boat services in 1946, it became necessary to re-establish Sourabaya as a refuelling stop between Darwin and Singapore. QEA's Captain H.T. Howse and Engineer George Lake arrived at Sourabaya in May and found it '... commercially a dead city. Damaged by enemy assault, neglected during occupation and battered by vicious Allied bombing, there remains only a nostalgic travesty of peace ...' (quoting Captain Howse). The town itself was devastated. It had been bombed, shelled, fought through and looted. The peacetime population of 250,000 was non-existent. Howse and Lake's major tasks were to prepare the harbour for commercial air use and organise point-to-point radio communication with Darwin and Singapore. By August 1946, flying boats alighting on Sourabaya harbour were met by a 40 foot patrol launch, a Shell Company refuelling barge and a marine passenger tender.The patrol launch, in addition to carrying out air-to-ground radio control, was used for sweeping the landing area. From touchdown to moorings, the flying boats taxied between sunken ships and bombed-out empty wharves. Two 15 cwt trucks, a sedan car and a jeep provided transport for the twelve Qantas Station staff at Sourabaya. The Dutch Civil Aviation Building, which had been used by Qantas before the war, was renovated as a flying boat base and office by QEA, and passengers were served morning tea in the comfortable surroundings of the refurbished building while the aircraft was refuelled.

In October 1945, QEA's Short S23 Flying Boat *Coriolanus* became the first civil aircraft to land at Singapore since February 1942. Under the command of Captain Keith Caldwell, the flight carried QEA's Superintendent Empire Operations Division, Captain Bill Crowther, who had joined the aircraft at Darwin, and Engineers Bill Bennett, Roy Carr, A.K. 'Kiwi' Stone, Norm Capel, Tony Koh, Tony Lopes and Bill Sykes. They were joined in Singapore by Captain Russell Tapp, who had flown there direct from Ceylon.

Their immediate objective, now that the war was officially over, was to investigate the possibility of re-routing the Australia–England Lancastrian mail plane service along the old Empire flying route through Singapore.

An official from the Australian Department of Civil Aviation, Mr R. Christie, who accompanied the Qantas team into Singapore, made preliminary arrangements to set up point-to-point radio communication with Darwin and also establish the weather broadcasting facilities. It was soon found that the old Seletar aerodrome, where the Qantas DH86s landed before the introduction of the Empire Flying Boats in 1938, had a landing strip that was not completely paved and would require considerable reconstruction to stand up to heavy traffic. Kallang aerodrome, which was still under the control of the RAF, had been greatly altered by the Japanese, and since it was built on reclaimed swamp land, it was considered that the Lancastrians and Liberators' heavy loads would probably break up the runway and drainage system, both of which had been recently repaired. However, the sea-plane anchorage at Kallang and its approach was judged safe for use by QEA and BOAC flying boats, following dredging of the area and reconstruction of the booms.

So it was Changi aerodrome, which had been built by Australian prisoners of war under Japanese control, that was finally opened for civil aviation on 8 April 1946, when a Liberator carrying passengers from Sydney to London became the first QEA service to resume operation through Singapore.

With the advent of the Constellations in December 1947, QEA's land-plane operation in Singapore was transferred to Tengah airport, since the runway at Changi was not considered adequate for the increased load of the new aircraft type. Tengah was about an hour's drive (18 miles) north-west of the city centre and had a full-scale concrete runway. Qantas and BOAC continued to use Tengah until 1950, when — following the appointment of a new Director of Civil Aviation at Singapore and a substantial upgrading of the runway at Kallang — heavy aircraft were given permission to operate from the enhanced airport facilities. From 1950 Kallang became the centre of activity for Qantas operations in Singapore.

When George Foord arrived in Singapore in 1948, having just completed his apprenticeship with Qantas, it was the largest company base outside Sydney. The Station was responsible for servicing QEA and BOAC aircraft operating between Sydney and London; BOAC aircraft on the London–Hong Kong route; QEA Lancastrians, which were now operating as freighters to the Far East; and QEA Hythe and BOAC Plymouth Flying Boats, which either passed through or terminated at Singapore. Bill Bennett had become QEA's Area Engineer, stationed in Singapore but with responsibilities extending to the flying boat base at Sourabaya and later to Djakarta. There were twenty-six Qantas engineers stationed in Singapore and more than 100 Singapore nationals. Jim Lamb was Senior Service Engineer and operated two distinctly separate service crews: an east-bound crew under Roy Carr, responsible for servicing all aircraft travelling from London to Sydney, from London to Hong Kong, or from London terminating at Singapore; and a west-bound crew under Ken Hopkins, responsible for aircraft heading in the opposite direction, including those from Sydney or Hong Kong that terminated in Singapore. George Foord was part of Ken Hopkins' west-bound crew. No matter what time of day or night an aircraft arrived at Singapore, the relevant crew would be called upon to carry out its servicing responsibilities.

The Singapore Line Station had its own Instrument Shop (with Neil Moncrieff and Doug Hamilton), an electrical overhaul shop (manned by 'Sandy' Armstrong and Darcy Dumbell), and a radio shop (with Jack Arthur, Cec Bennett and Johnny Williams). When a component failed during flight, it was removed from the aircraft at Singapore and taken to one of the engineering workshops for repair or even overhaul, before being returned to the aircraft for its scheduled 7 am departure.

George Dusting and Roy Kendall were both sub-foremen in Singapore at this time, and other engineering staff included Jack Craik, Arthur Kemp, Jack Gould, Merv Marsh, Bill Hodge, Ken Hodgkinson, Graham Marriott, Laurie Curley, Ernie Darking, Jack Rowe, Jack Reid and 'Bluey' Hammond. The QEA engineers at Singapore developed an enviable reputation for their efficient handling of aircraft and the manner in which they enabled aircrews to maintain flight schedules. According to George Foord,

> 'Although BOAC had Stations at Calcutta, Karachi, Cairo, etc., the same degree of engineering expertise was not considered to be there. The QEA aircrews would have their aircraft serviced in London and then attempt to get to Singapore without having to have repairs carried out along the way. Once they got to Singapore, they were as good as home. They all relied a lot on us and respected Singapore as a Station. Not only was the ability there, but the engineers at Singapore also had the right attitude — the blokes were there to make it work.
>
> It was different in BOAC Stations. Our aircrews often complained of having a group of BOAC engineers working on a QEA aircraft when a BOAC service would arrive. They would immediately down tools and leave the Qantas plane to work on their own Company aircraft. That just didn't happen at our Station in Singapore.'

In 1950 Bill Bennett left Singapore and returned to Sydney where he was subsequently promoted to the position of Superintendent Route Stations–Engineering in 1952. He was replaced as Singapore Area Engineer by Roy Carr.

When Qantas began operating Lancastrian freighters and passenger-carrying DC4 Skymasters through Darwin, Djakarta and Manila to Japan in 1949, Bluey Hammond was sent from Singapore to Djakarta to supervise the servicing of the QEA aircraft (which was being undertaken by KLM) because he had licences for both aircraft types and was able to certify them on the company's behalf. Djakarta eventually became a QEA Line Station with George Foord as Station Engineer.

In September 1951 George Dusting was appointed Station Engineer at Manila, replacing Dick Galloway who had been QEA's engineering representative in the Philippines since the company began its civilian flights to Japan in 1949. Then when George Dusting was sent to England in March 1953 for training on the De Havilland Comet, Bill Barrie was made Acting Station Engineer–Manila, a position he held for just on twelve months.

During the period George Dusting was away, BOAC introduced a twice-weekly Comet service to Tokyo through Manila, so Bill Bennett flew to the Philippines to organise the handling of jet aircraft at the QEA Line Station. Neither the Acting Station Engineer, Bill Barrie, nor engineer Don White were licenced on the Comet, but they carried out the servicing under the supervision of BOAC's engineer, Jack Ellis, who was seconded to Qantas pending George Dusting's return. He was licenced on the new aircraft type and therefore responsible for its certification at the completion of each transit service. (Comets had begun flying from London to Singapore in October 1952, and Ken Hopkins, who by this time was QEA's Area Engineer at Singapore, had undergone a nine-week Comet training course in England, receiving a British Air Ministry 'A' licence on the aircraft. This enabled him to sign the Comet out after its turn-around service in Singapore without the necessity of having a BOAC engineer present.)

Upon his return to Sydney Bill Bennett reported on QEA's handling of the Comet at Manila:

> 'Time is the most vital element in 'Comet' handling since BOAC recently lowered their scheduled time on the ground, from chock to chock, for the 'Comet' at Manila, from one hour to forty minutes. This means that every move made by engineers, traffic and ground staff must be pre-planned and that extensive use must be made of aircraft handling equipment.
>
> Two Dorset refuellers move into positions to the port and starboard wing tips. A tow-motor brings into position a power unit, a baggage trolley containing undercarriage jacks, two main wheels and tyres made up complete, and a nosewheel and jacks. It also carries two buckets of water containing Teepol. During refuelling, it is most essential that everyone working on the plane should regularly wash their hands in water and Teepol, as turbine fuel can cause dermatitis very easily.
>
> The engineer in charge checks the undercarriage tyres and the two filling points, reading the dipstick in preparation for refuelling. If a wheel has to be changed, fifteen minutes is allowed for this, as it must be done before refuelling. The fuelling is

commenced with anything up to 6000 gallons being taken on board. Time allowed is twenty minutes. The aircraft visibly sinks down as fuel is taken on board.

A check is made on oxygen and compressed air needs, an engine check is made, and the oil topped up.

The power unit supplies the 1200 amps necessary to start the engines. The engineer in charge has telephone contact with the aircraft Captain. Everyone near the 'Comet' wears ear muffs because of the tremendous noise made by the jet engines.'

George Dusting returned to Manila in March 1954 as the first Australian to hold Series 1 and Series 2 Comet and Ghost engine endorsements on his DCA licence. He was referred to at the time as 'Australia's First Jet Engineer'. Then in August 1954 he became QEA's Station Engineer at Singapore, replacing Ken Hopkins, who had held the position since 1952. George Dusting's position as Station Engineer in Manila was taken over by Alan Keats.

When QEA began its Constellation service across the Indian Ocean to Johannesburg in South Africa in September 1954, Engineering Line Stations were established along the route at Cocos Island, where Ken Barnes was appointed Station Engineer in March 1953, and at Mauritius, which was manned by Billy Monds and Joe Wiggins. Cocos Island became an important junction in QEA operations when, simultaneous with the opening of the South African route, the company introduced a DC4 Skymaster service that connected with the new Constellation service from Johannesburg at Cocos Island, providing a direct link for passengers between South Africa and the Far East. The DC4 link flew from Cocos Island to Japan, via Djakarta, Manila and Hong Kong.

The opening of the Southern Cross route across the Pacific in May 1954 with the company's new fleet of L1049 Super Constellations saw a need for the establishment of an additional Engineering Line Station at Nadi (Fiji), and QEA engineering representation at Canton Island, Honolulu and San Francisco. Qantas had taken over the Pacific operation from BCPA, which had been flying Douglas DC6s on the route. TEAL bought the DC6s from BCPA and began operating a feeder service between Sydney, Auckland and Nadi, since the QEA service no longer flew through New Zealand.

After extensive training on the Super Constellations at Lockheed in the United States, George Foord went to Nadi as QEA's Station Engineer, accompanied by engineer Charlie Eames.They arrived ten days before the start of the QEA service. George Foord describes the scene that confronted them:

'BCPA had been operating from an old hangar workshop, which was too small to take an aircraft, and it was in a disgraceful condition. The roof had leaked for so long that the floor was rotten. What few spares they had were just stored in boxes, not racked

> or catalogued. The BCPA blokes knew little about the airframes, electrics, air conditioning or auto pilots on the aircraft. The whole situation was just a disgrace from our point of view — I don't know how they coped.'

Nadi airport was under the administrative control of the New Zealand Air Department, and George Foord arranged through them to have the roof fixed on the hangar workshop. He then set about constructing some suitable work benches and racks for the storage of spares. Initially they lived in grass huts next to the airstrip, about 250 yards from the runway, but Qantas eventually built proper houses for use by their staff who were stationed in Fiji.

The acquisition of suitable equipment for operating the Nadi Station proved difficult in the early days. BCPA had an old truck, similar to the English Austin army ambulances used during the war, which was more than ten years old when it was taken over by Qantas. George Foord made several attempts to get permission to buy a more suitable vehicle for use by the engineers, but to no avail. No-one in Sydney seemed to realise just how desperate the situation in Nadi was until one day, Captain Bert Ritchie, who was QEA Technical Manager at the time, passed through Fiji on his way to the United States. George Foord took the opportunity to demonstrate to Captain Ritchie just how dilapidated the old truck actually was, and within a very short time the engineers at Nadi were driving around in a nice new 1954 model FJ Holden Utility, which was fitted with a canopy.

Charlie Eames stayed at Nadi for about five months and was replaced by Brian Hendy, who had just completed his apprenticeship and subsequently became a Qantas Flight Engineer. Then when Stan Quigg joined the QEA engineering team at Nadi, the two original BCPA engineers became redundant and their services were terminated.

Qantas was unable to establish its own Engineering Line Stations at Canton Island, Honolulu, San Francisco and Vancouver initially. Apart from the fact that it was not an economical proposition, with a frequency of just two flights weekly operating between Sydney and San Francisco, and three of those flights extending to Vancouver each fortnight, the QEA engineers were unable to get permits to allow them to work in the United States or American-controlled territories. So QEA entered into agreements which saw America's Pan American Airways service the Qantas Super Constellations at Canton Island, and United Airlines carry out the aircraft servicing function at both Honolulu and San Francisco. Refuelling at Honolulu was handled by the Shell Company.

Qantas was, however, able to have its ground engineers, who were licensed on the Super Constellations, stationed at these locations in a strictly supervisory role as Engineering Liaison Officers to the American airline doing the actual servicing work. They were permitted to enter the United States on B1-B2 visas which, by the letter of the law, allowed them only to oversee the servicing procedures and certify the aircraft.

Canton Island, a small coral atoll covering an area of only 6 by 3 miles, is situated in the mid-Pacific between Fiji and Hawaii. It was a base of great strategic importance to the Americans during the war, and when Pan-Am established its own civil station on the island in the late 1940s, all the wartime facilities were still in place. It was therefore well equipped for a Line Station operation. Qantas used Canton Island as a technical port of call for refuelling and checking vital weather information. But if an aircraft was delayed for an extended period, barrack-type accommodation was made available for use by stranded passengers. Harold Avery was QEA's Engineering Liaison Officer on Canton Island, but he remained there for only six months and was replaced in October 1954 by Jeff Harris, who had undergone an additional period of training on the Super Constellations at Lockheed.

Lindsay ('Sandy') Armstrong became QEA's Engineering Liaison Officer at Honolulu, while Ken Barnes was posted to San Francisco. On route to take up his post at San Francisco, Ken Barnes found himself in strife with the American authorities. Travelling on a B1-B2 visa, he arrived at Honolulu to be asked by US Immigration officials why he was travelling to America.

'I'm going to work there,' was Ken's reported reply.

'Are you now?' came the official's response, as he continued to examine Ken's visa. 'Would you mind stepping into this room for a moment, sir, while we clear up this matter.'

Ken entered the room, and the door was closed and locked. He was virtually placed under 'house' arrest overnight for the more than twelve hours it took Qantas to clarify his position. The matter resolved, he was allowed to continue his journey to San Francisco. But word passed quickly down the line to the other engineers as to the correct line of approach to take with the US Immigration officials.

At Vancouver the turn-around servicing of Qantas aircraft was undertaken by Trans Canada Airlines. Both Roy Carr and Barry Watham spent an initial three months in Vancouver as Engineering Liaison Officers. But the economics of keeping resident engineers in Vancouver to supervise the servicing of three flights per fortnight soon proved unviable, so after resident Qantas engineering representation at Vancouver was withdrawn, Ken Barnes began accompanying the QEA services north from San Francisco to carry out the supervisory function at the Canadian route terminus.

As Qantas took delivery of additional Super Constellations and increased the frequency of its services across the Pacific, the need for the company to establish fully staffed QEA Engineering Line Stations at Canton Island, Honolulu and San Francisco grew. The main difficulty was that Australians were unable to get 'Green Cards' permitting them to perform work in the United States that could be carried out by similarly qualified American citizens. The trade unions eventually agreed to allow a certain level of Australian engineering representation, but those people involved were to work in supervisory roles only. The ground staff employed in the physical handling of aircraft had to be American engineers.

By mid-1959 Ken Barnes had moved to Canton Island, and Sandy Armstrong had become Area Engineer–North America, stationed in San Francisco. The Qantas Line Station at San Francisco had a staff of 100, thirty-six of whom were engineering personnel. It became the company's North American headquarters, and the ground staff were handling five QEA Super Constellation services per week, in and out, between the American west coast and Australia, one in and out to Vancouver, and two in and out to London. In addition, two BOAC services from London were turned around each week at San Francisco, and they too were handled by Qantas staff.

By the beginning of 1960, after the introduction of Qantas and BOAC jet services into North America, the QEA Line Station at San Francisco was handling twenty-two in and out services a week, and its staff numbers had swelled to 140.

In September 1955, 'George' A.G.S. Pandian became the first local Singapore staff member to gain a Grade 1 Ground Engineer's licence under the Qantas Training Scheme. George had joined Qantas in 1949 as a trainee engineer, and was sent to Australia in 1952 for training at the QEA Technical School. On graduation he returned to Singapore as a fully fledged ground engineer, certified to take over maintenance of the L749 Constellations.

Paya Lebar replaced Kalang as the main civil airport in Singapore in August 1955. The strip at Paya Lebar was directly aligned with the runway at Kalang, and on the day of the official opening of the new airport, Qantas recorded its shortest scheduled flight in a Super Constellation. Under the command of Captain Ralph Bruce, the aircraft took off from Kalang and touched down fifty-two seconds later at Paya Lebar. There was not sufficient time for the Captain to raise and lower the landing gear during the flight. The QEA Line Station closed at Kalang on the evening of Saturday 20 August 1955 and resumed operation the following morning at Paya Lebar.

In 1959 Bill Bennett was promoted to Chief Stations Engineer, a position he held until 1967 when his deputy, Kit Brown succeeded him and became Line Maintenance Manager, stationed in Sydney. With the rapid expansion of Qantas services and the resultant establishment of additional Engineering Line Stations at strategic locations around the world, the administration of the division was split in the mid-1960s and placed under the authority of three Line Maintenance Superintendents, subordinate to the Line Maintenance Manager.

The Line Maintenance Superintendent–Eastern Division was responsible for all Qantas aircraft maintenance and associated contracts for maintenance work at all ports of call from Indonesia westward through Singapore and Europe to the United Kingdom, as well as Cocos Island, Mauritius and Johannesburg. The Superintendent-in-charge of the Eastern Division covered Stations where QEA engineers were posted in Tahiti, Nadi and across the Pacific to the Americas, which by this time included Mexico City, Acapulco and Bermuda. (Servicing of QEA aircraft passing through New York was handled by BOAC.) The third Superintendent, whose responsibility was the Central Division, controlled the engineering operations in Australia at Darwin, Brisbane, Melbourne and Perth, as well as those in Japan, Hong Kong, the Philippines, New Guinea and New Zealand.

Doug Hamilton had transferred to Singapore from Rose Bay in early 1948 to work on the repair and overhaul of aircraft instrument systems. When Neil Moncrieff resigned from Qantas to join Malayan Airways at Kalang airport in 1951, Doug Hamilton was asked to take over the responsibility of managing the QEA Instrument Shop at Singapore. Then in 1954, with the introduction of the L1049 Super Constellations on the Kangaroo route, the Instrument and Electrical Shops were incorporated into one department, and the combined facility was placed under Doug Hamilton's control.

He obtained British ARB licences on the L749 Constellation airframe and engines in 1957, and then transferred to Sydney to undergo a four-month course on the L1049 Super Constellations, where he gained endorsements to his DCA licence for both engines and airframes on that aircraft. In January 1958, he was positioned at Manila as Maintenance Sub-Foreman.

At the time, the QEA Line Station at Manila was a seven-day-a-week operation handling Qantas L1049 Super Constellations and DC4 Freighters; DC3s, DC4s and DC6s for Cathay Pacific; Comet 4s and Argonauts (a British-engined derivative of the Douglas DC4, built in Canada by Canadair) for BOAC; and Vickers Viscounts for Hong Kong Airways. In 1959 Cathay Pacific purchased two Lockheed L188 Electras and introduced them onto their Hong Kong–Manila turnaround services as well as their Hong Kong–Manila–Darwin–Sydney route. (Qantas began operating its Electras to the Far East in December 1959.) But no-one at QEA's Line Station in Manila was licensed on the L188s, so Doug

Hamilton was sent to Lockheed in February 1959 to do the Electra and Allison 501 engine courses, and gain the necessary licence endorsements for maintenance and certification of aircraft type passing through Manila.

In February 1961, he returned to Sydney again, this time to attend a six-month training course on the Boeing 707s and the Pratt and Whitney JT3 'C' and 'D' series engines. Now licensed to certify the maintenance and servicing of the Qantas 707s, he was posted to Nadi in November 1961 as Station Engineer.

Qantas was one of three airlines operating through Fiji between Australia and the United States and Canada, the other two being Pan American World Airways (Pan-Am) and Canadian Pacific Airlines. Pan-Am was operating a daily Boeing 707 service to and from Australia through Nadi, and a daily DC6 feeder service to New Zealand.

Canadian Pacific was flying turbo-prop Bristol Britannias on a fortnightly basis across the Pacific to Sydney, while Qantas was handling seventeen of its own 707 flights through Nadi each week — two services a day between Australia and the United States (one east and one west) and three turn-around flights a week from either Sydney or Brisbane. In addition, the QEA engineers were servicing Qantas and TEAL Electra feeder services through New Zealand, which the two airlines were operating in partnership under a pool arrangement. (Initially, both Pan-Am and Canadian Pacific had their own engineering personnel stationed at Nadi to service their aircraft as they passed through Fiji).

By 1964, the original old workshop/hangar at Nadi had been replaced by a two-storey building, which was occupied solely by Qantas Station personnel. The upper level contained the offices for Management, the Accounts Department, the Property Officer and sales staff, while the ground floor was used for stores, catering, an engineering workshop and a large undercover area, like a small hangar, for equipment parking and aircraft maintenance.

Qantas had already taken over the servicing of Canadian Pacific's aircraft on the Pacific route when, in 1964, Pan-Am also withdrew its maintenance personnel from Fiji and contracted with Qantas for the servicing of its aircraft, leaving one of their Engineering Liaison Officers on hand to certify the Pan-Am aircraft before flight.

In January 1967, Doug Hamilton returned to Singapore as Station Engineer. Then in July 1968 he was appointed Line Maintenance Superintendent–Western Division, stationed in Sydney, a position he relinquished in 1974 to take over similar responsibilities as the Eastern Division Superintendent.

Meanwhile, Kit Brown had been replaced as Line Maintenance Manager by John Mills and then by Leo McCarthy. It was the direct responsibility of the Line Maintenance Manager to negotiate all contracts covering engineering facilities and maintenance for Qantas aircraft at all ports of call, other than the company's main Jet Base in Sydney. He also drew up agreements with customer airlines for maintenance of their aircraft at Stations where Qantas had the prime responsibility for providing them with appropriate services and equipment. When Leo McCarthy retired in 1979, Doug Hamilton was appointed to replace him as Line Maintenance Manager. At that time, Qantas had either full engineering facilities or engineering representation to certify the company's aircraft at thirty-eight Line Stations around the world.

CHAPTER

16

CONTINUING THE TRADITION

In preparation for the introduction of its new jet aircraft onto the Southern Cross route to San Francisco and the Kangaroo Service to London, Qantas established a Boeing 707 Jet Training Centre at Avalon, near Geelong in southern Victoria. The facility was not only for the instruction of flight crews on the 707s, but was also used extensively for training ground engineers and maintenance personnel who had not had the opportunity to attend courses on the new aircraft type at Boeing in the United States.

Norm Roberts was appointed Manager of the base operation at Avalon, and he was responsible for every person who entered the Qantas area. Because the Qantas 707s were operating from the Government Aircraft Factory strip in a 'security classified' area, all Qantas personnel and visitors to the Training Centre had to have security clearances.

Don Smith was Station Engineer – Avalon, and working with him in the on-the-job training of the Qantas Engineering ground staff were Boeing Service Engineers Tom Deschner and Al Green, together with Pratt and Whitney's John Lucas. Arthur Rowe, who was Superintendent Flight Engineering (Training), had four engineering instructors in his team: George Riek, Ted Fargas, Keith Vickerman and Boeing's Ken Longmuir. Representatives of both Shell and AWA were also positioned at Avalon.

The original plan was to have 120 pilots and forty-two flight engineers trained on the 707s by the end of October 1959 to man the Qantas round-the-world operations, so the training at Avalon was intense. In one day in August, two 707s recorded nineteen landings and sixty-five touch and goes. Under the guidance of experienced ground engineers, trainee ground staff were able to refuel an aircraft and change a wheel in thirty-five minutes. The jets kept flying in all weather, rain, hail or shine. Only a fierce crosswind saw them grounded.

During this period, Norm Roberts was living at Carss Park, a waterside suburb in Southern Sydney. Each morning he would leave home at 6 am and drive to Mascot, where he would buy a copy of the Sydney Morning Herald, board a Qantas 707 and spend the next hour flying to Avalon for the day's work. At the end of the day he would reverse the procedure, this time buying the Melbourne Age newspaper to read during the afternoon flight to Sydney. He would usually arrive home at about 6 pm — not a particularly long day in Qantas Engineering terms — but few people could claim to have travelled in excess of 1200 miles each day just to get to and from work.

By the end of October 1959, Qantas had begun operating its 707s on the Kangaroo route to London, linking with the service via the United States to provide the first round-the-world jet service through New York. Then, during November and December 1959, Qantas took delivery of its four Lockheed 188C Electras (VH-ECA, named Pacific Electra, VH-ECB Pacific Explorer, VH-ECC Pacific Endeavour and VH-ECD Pacific Enterprise), and immediately placed them into service on the Far East routes to Hong Kong (through Port Moresby and Manila) and Japan (through Darwin, Manila and Hong Kong). The Electra could carry 'a minimum of forty-two deluxe passengers in sleeper chairs, or a maximum of seventy-seven tourists' and a crew of nine. Its cruising speed of 400 miles per hour 'straight and level' at 20,000 feet made it the fastest turbo-prop aircraft available at the time.

Bob Walker was QEA's Controller of the Electra project, while the company's Resident Engineer at Lockheed during the assembly of the Qantas aircraft was Phil Jost. As the delivery dates for the Electras drew nearer, a number of Qantas engineers attended courses in the United States on the new aircraft. Those known to have received training at

Burbank included: Darcy Dumbell, Bob Morton, Neville Burton, John Hewitt, Doug Hales and Alan Hansen, who all completed an electrical and instrument maintenance course; Don Hollebon, Rolly Hill, Alan Woods, Keith Hancock, Gordon Hall and Bruce Scott, an airframe and engine maintenance course; and Keith Ower, Jack Doldissen, Harold Avery, George Prasser and Noel Manning, an airframe engineering course. Production Inspectors Bill Thomas, Dave Cary, Harry Wright, Frank Ward and Norm Heppell also went to Lockheed for instruction on the Electras, returning to Australia on the delivery flights of the second, third and fourth aircraft. Bob De Coque, who was to be in charge of servicing of the Electras at Sydney Airport's International Terminal, attended a general aeroplane maintenance course at Burbank and then spent two weeks on the production flight line studying maintenance and handling problems.

An additional twenty-four engineers underwent practical training on the Electras at Avalon during the first two months of 1960, after the aircraft had entered service on the Qantas Far East routes.

Following the introduction of the Electras, Qantas began to accept contract work from other airlines and governments for the overhaul of Allison engines. In the first twelve months, 184 Allison engines passed through the new Qantas engineering facilities at Mascot: thirty for Ansett-ANA, forty-three for TAA, fifty-three for the RAAF and fifty-eight of Qantas' own engines. Then in October 1960, Qantas was awarded a substantial overseas contract (worth £200,700 per year) to overhaul Allison engines for Garuda Indonesia Airways, who were planning to introduce Electras onto their domestic and international routes in January 1961. The agreement included the proviso that Qantas help train Garuda's technical people in the handling of Electra aircraft, and five of the airline's supervisory engineering staff (Inspectors Tgeng Biauw Kiat, Lie Kim Gam and Suparto, and Crew Chiefs Legimo and Zainul Arifin) and the Chief of their Engine Overhaul Department, P. Muljono, all came to Sydney to benefit from QEA's twelve months' experience with the new aircraft type. The contract was negotiated by QEA's Engineering Contract Administrator, Harold Hutton, and finalised in Djarkata by Qantas' Manager for Indonesia, John Hill.

Qantas celebrated its fortieth anniversary in 1960. To mark the occasion a replica of the company's 1922 booking office in Longreach was put on public display in the area opposite the new Qantas House building, which is now known as Chifley Square. And the impressive Qantas Fortieth Anniversary symbol, measuring 10 feet in diameter, was emblazoned across the tails of the company's fleet of 707s.

According to Qantas Engineering Manager, Doug Hudson, the efficiency and reliability of the Pratt and Whitney JT3C-6 engines on the Boeing 707s caught the Qantas engineers by surprise, prompting a re-examination of the company's maintenance philosophy. The JT3C-6 was a 'developed' engine by the time it was chosen to power the 707s, as it had been in operation on US military aircraft since the early 1950s. Although the Qantas engineers recognised that the turbo jet engines operated on a completely different principle to the reciprocating Wright Cyclone turbo-compound engines on the Super Constellations and were therefore subject to different stresses, the JT3C-6s were still programmed for maintenance initially at the same frequency as the piston engines. But it soon became apparent that the prescribed 400 hours operating time between overhauls was unwarranted, and within twelve months that period had increased to 2200 hours.

The Pratt and Whitney engines on the Boeing 707s were pure jets. Put simply, *all* the air drawn into the front of the engine was compressed by rotating compressor blades and fed into combustion chambers, where the fuel was ignited and then passed at high velocity from the rear of the engine. Apart from the fact that the engines were very noisy, the process did not provide a desirable amount of thrust, because relatively small volumes of air passed through the narrow engine at very high speed.

Within months of the introduction of the Qantas 707-138s, new jet technology became available, capable of passing larger volumes of air through the engine at lower speeds. It was called the Fan Jet, since some of the first stages of the compressor blades were lengthened to make a mini-propeller inside the cowling. A proportion of the air entering the front of the engine now bypassed the combustion process. It moved around the main engine core and mixed with the hot exhaust effluent to pass rearward. These engines, designated the Pratt and Whitney JT3D-1, not only provided a substantial increase in thrust (from 13,500 to 17,000 pounds) but markedly reduced fuel consumption. Qantas was quick to recognise the benefits of the Fan Jet engine, arranging for them to be fitted to their four new 707s ordered for delivery in 1961. They also embarked on a major program to modify the engines on their existing fleet of seven 707s, which were progressively returned to Boeing where the 707-138 airframes were modified to 707-138B specifications, while twenty-nine engines were converted from JT3C-6s to the new Fan Jet JT3D-1(MC6) specification by American Airlines. (Note: The designation MC6 was applied only to engines that were *converted* to Fan Jets from JT3C-6 specifications.)

Meanwhile, at the Qantas Repair Overhaul Base (as it was now called) at Mascot, company engineers converted a further twenty-five engines, which had been brought in from Line Stations, to turbofan specifications, using a series of 'quick engine exchange kits' provided by Pratt and Whitney.

In August 1961 Qantas took delivery of it first four Boeing 707-138Bs

(VH-EBH, named *City of Darwin*, VH-EBI *Winton*, VH-EBJ *Longreach* and VH-EBK *City of Newcastle*). Shortly afterwards, the company introduced the designation V-Jet to all eleven of its 707 aircraft, signifying that its turbofan fleet was superior in performance to aircraft of all other international airlines serving the Australian market. The term 'V-Jet' came from the Latin word *vannus*, which means 'Fan'.

In August 1960 Lockheed initiated a $25 million recall program of its turboprop Electras after a crash in California's Sierra Nevada mountain range, in which all on board were killed. Speed restrictions were immediately imposed by the American Federal Aviation Authority on all Electras flying in the United States. Following the American lead, Australia's Department of Civil Aviation also imposed speed restrictions on the Electras flying in this country. As it happened, these speed restrictions had no effect on the four Qantas Electras as they were already operating at the reduced speed for economic reasons, in accordance with the recommendations of the company's Performance Engineers. They were, however, returned progressively for modification under the Lockheed Electra Action Program. Tests had revealed that a weakness existed in the mounts holding the engines in their nacelles. At a certain speed and under certain conditions, vibration caused severe strain on the mounts, allowing the engine to become unstable on the wing. This caused a chain reaction, which culminated in a flutter so violent that the wing separated from the fuselage.

After their return to Australia, the modified QEA Electras were introduced on the trans-Tasman services when Qantas began operating the route to New Zealand in its own right in October 1961.

It took until February 1962 for the new Line Maintenance Hangar and workshop annex to come into operation. (Until then, maintenance on the 707s had been carried out in the open-fronted Hangar 96 and from special scaffolding which was built outside on the airport apron.) Built at a cost of £760,000 ($1,520,000), the Hangar 131 complex was planned to be more flexible than Hangar 96, which had a permanently fixed centre column. The main front truss was designed so that the supporting columns could be placed at two alternative positions: in the centre of the span to divide the hangar into two equal bays to accommodate two Boeing 707s or relocated to divide it into three bays of equal size to house three Electras simultaneously. (As it happened, Hangar 131 was never used for the maintenance of Electras and remained permanently a two-bay hangar for the Boeing 707s.)

Unlike the open-fronted Hangar 96, Hangar 131 was sheltered from the elements by six 55 feet wide by 49 feet high front doors, each weighing about 18 tons. Electronically operated, the steel-frame, aluminium-clad doors were fitted with green corrugated-fibreglass panels to provide the hangar interior with additional lighting. Structural Engineers for the hangar and the workshop/office complex were McDonald, Waggner and Priddle, who worked in consultation with Professor J.W. Roderick, Dean of the Faculty of Engineering at the University of Sydney.

Construction of the Hangar 131 complex had begun in late 1960, following the diversion of the Sydenham–Botany railway goods line which, until that time, had run along an embankment just north of the existing hangar line and had actually crossed the then north–south Runway 04-22 of the airport, about 500 feet from the northern end of the strip. (On the evening of Sunday 18 June 1950 at 7:59 pm, an Ansett Airways DC3, while preparing for takeoff to Brisbane via Coffs Harbour with fifteen passengers and a crew of three on board, collided with an empty coal train from Bunnerong Power Station, which was crossing the runway at the time. Although the aircraft and three of the train's coal trucks suffered substantial damage, remarkably there was no loss of human life in the accident.)

By 1964 the term 'foreign order' had taken on a new meaning for Qantas Engineering. Outside contract work had become an important revenue earner for the company, and Qantas began to seek more markets for its engineering resources and technical expertise in aircraft and engine types not in its own fleet.

A special Engineering Sales and Service Department was established under Engineering Sales Controller Harold Hutton, to service the engineering requirements of the company's outside customers. Barry Marshall was appointed Customer Service Officer (Overhaul), and Peter Drew became Customer Service Officer (Maintenance). Their function was to ensure that work being undertaken for the customer was carried out in accordance with the terms of the contract. Revenue from outside contract work was running at £1.5 million ($3 million) per year, and customers included the Royal Australian Air Force (RAAF), Royal Australian Navy (RAN), Trans-Australia Airlines (TAA), Tasman Empire Airways (TEAL), South African Airways, Malaysian Airways, Cathay Pacific Airways, Garuda Indonesian Airways, Pakistan International Airlines and Air India. In the twelve months to September 1964, Qantas had overhauled 172 engines and more than 5000 components for customer airlines at its Jet Base engineering facilities at Mascot.

The rapid development of the outside contract business reflected the comprehensive and high standard of engineering facilities that Qantas had always maintained. This, of course, had been dictated by what was known as 'the 10,000 mile factor' — the geographical remoteness of the Qantas base of operations from the manufacturers of the company's aircraft, engines and components. Qantas always had to be equipped, from its own resources, to service every component from each aircraft type flown by the company, and an increasing number of overseas airlines came to rely on the engineering excellence that Qantas had developed and now had to offer.

In August and September 1964 QEA placed its last two 707-138Bs (VH-EBL, named *City of Geelong*, and VH-EBM, named *City of Launceston*) into service on its round-the-world routes. Then in February 1965, Qantas took delivery of what was to be the first of twenty-one Boeing 707-338Cs (VH-EBN, named *City of Parramatta*) purchased by the company. Although the 338Cs were 17 feet longer than the 138Bs, the passenger configuration remained at the existing twenty First Class and eighty-four Economy Class, the extra space being reserved for cargo capacity of an additional five and a half tons. They were powered by four Pratt and Whitney JT3D-3B engines, each of which provided 18,000 pounds thrust.

On 8 March 1965, the second Qantas 707-338C (VH-EBO, named *City of Townsville*) became the first aircraft to cross the Pacific non-stop from San Francisco to Sydney. It was the longest flight undertaken by a civil aircraft, the 7500 mile journey being achieved in 14 hours 32 minutes. The arrival of the new aircraft and an increase in Qantas services led to a further expansion of the engineering facilities at Mascot. A second line maintenance hangar, Hangar 191, was built on the eastern side directly adjacent to Hangar 131, and Hangar 96 (which was being used for block overhauls) was expanded 25 feet in depth and fitted with two 50 foot front doors to enable it to handle the additional length of the 707-338Cs.

Qantas Chief Projects Engineer Bob Walker tranferred to London in early 1965 to become the company's Technical Representative for the United Kingdom and Europe. His main responsibility was to liaise with Hawker Siddeley, manufacturer of the HS-125 mini-jets (two of which Qantas had on order for its flight training program) and the suppliers of instruments and hydraulic equipment for the aircraft. He was also to keep in touch with developments in the Anglo-French Concorde project since Qantas, at that stage, had four of the supersonic aircraft on order.

Tasman Empire Airways Limited (TEAL) changed its name to Air New Zealand in April 1965, just as Qantas began operating its 707 V-Jets across the Tasman between Sydney and Christchurch. The Qantas Electras continued to fly from Sydney to Auckland and Wellington, although Qantas did sell one of its Electras (VH-ECC) to Air New Zealand, to replace one lost by the airline in a training accident at Auckland.

May 1965 saw BOAC withdraw its Comet 4s from the Kangaroo route and replace them with Boeing 707s.

In mid-1966, the two Hawker Siddeley HS-125 series B jet trainers (VH-ECE and VH-ECF, both unnamed) were delivered to Qantas. Fitted out as 'baby Boeings', the HS-125s had instruments, systems and internal equipment similar to the Boeing 707s, and the usual 'executive' interior was modified for air crew training. They served as airborne simulators for the Qantas Boeings, allowing pilots to build up jet command experience at a fraction of the cost of flying the 707s. With a cruising speed of 430 miles per hour and a range of 800 miles, the HS-125s were an ideal training aircraft. They were powered by two Bristol Siddeley Viper 522 jet engines mounted at the rear of the fuselage and had swept-back wings similar to the Boeing 707s. To ensure that the overhaul sections at Mascot were geared to service the HS-125's Viper engines, a group of Qantas engineers had gone to England during 1965, spending a week at Bristol Siddeley's plant at Bristol before moving on to Coventry to study all aspects of tooling requirements, special processes and planning. Members of the group were Eric Walsh (Superintendent – Power Plants), Bruce Bounds (Sub-foreman – Engine Repair Section), Wally Green (Planning Supervisor – Power Plants, ROS) and Ray Richards (Procurement and Supply Department). They were joined later by Lee Charles (Technical Services), who assessed the nature of modifications necessary to the Jet Test Cell at Mascot to allow for the testing the Viper engines after each overhaul.

In June 1966, Sir Hudson Fysh retired as Chairman of the Qantas Board. His career had spanned the entire forty-five year history of the company, and he had played a leading part in its development and expansion. One of the many farewell luncheons for Sir Hudson was held in the 707 training mock-up at Mascot and was attended by the Engineering Department's 'senior citizens'; those seated at the function represented 504 years of service to Qantas. Sir Hudson Fysh was succeeded as Chairman by Sir Roland Wilson, who was at that time Secretary to the Commonwealth Treasury and a Director on the Qantas Board.

Within twelve months of Sir Hudson Fysh's retirement, Chief Executive and General Manager C.O. Turner had also retired. In the Queen's Birthday Honours for 1967, his imagination, foresight and devotion to Qantas was recognised with the awarding of a Knighthood. Sir Cedric's deputy, Captain R.J. (Bert) Ritchie, became Chief Executive and General Manager in July 1967. In the management reorganisation that ensued, Ron Yates was appointed to the new position of Director of Technical Services, Bob Walker became Technical Services Manager, and Doug Hudson retained his title of Engineering Manager.

The contribution of the Qantas engineers in the development of the company to its international status was also recognised when New Year's Honours lists in the 1960s included MBEs (Member of the British Empire) for the company's first three apprentices, Chief Maintenance Engineer Eric Kydd, Engineering Facilities Controller Jack Avery, and Chief Stations Engineer Bill Bennett. The awards were made for Services to Civil Aviation. Chief Engineer, Ern Aldis was honoured with an OBE (Order of the British Empire).

On 1 August 1967, the word 'Empire' was dropped from the name 'Qantas Empire Airways Limited', and the company became known simply as Qantas Airways Limited.

By the early 1960s, Qantas had an apprenticeship strength of 320, and its Apprentice Training Scheme had gained an enviable reputation of excellence in international airline circles. The Licenced Aircraft Maintenance Engineers in Qantas, who had trained through the Qantas Apprenticeship system, were regarded as the highest standard of airline engineer to be found anywhere in the world. In 1966 the Apprentice Training Centre was moved from the Jet Base at Mascot to a four-storey building at nearby Rosebery. It became one of the largest and most sophisticated establishments of its type in Australia, and was equipped with machinery worth $164,000, the installation of which was undertaken mainly by second-year apprentices. A staff of forty-one catered for an annual intake of 220 engineering apprentices. Jan Aldous had been appointed apprentice training controller and was one of the principals behind the move to Rosebery. According to Jan Aldous ...

> 'The aim of the Apprentice Training Scheme operating within Qantas at that time was to help provide the Company with the numbers of skilled tradesmen required to maintain its aircraft fleet and associated equipment. Apprentices were recruited in seven major trade categories (i.e. Airframe, Engine, Electrical, Instrument, Radio, Sheet Metal and Fitting and Machining), but a few were also trained as Welders, Electroplaters and Simulator Technicians.
>
> We had five classrooms and eight workshops at Rosebery, each equipped with the most up-to-date workshop and classroom equipment available. The Basic Engine Shop had ten workbenches, complete with two vices each, and ten Volkswagen engines. We used these for two reasons — because good quality engines and spares were readily and cheaply available from the local VW wrecker, and also because they were similar to engines that were being used on some light aircraft. The boys built a test bed capable of running an engine, so that any of the engines could be test run at the end of each session.
>
> The Gas Turbine Shop also had ten workbenches with two vices each, and four Armstrong Siddeley Mamba gas turbine engines. We got these from the Royal Australian Navy in good condition and complete with log books. Being axial flow engines, they closely resembled the Pratt and Whitney engines on the Boeing 707 aircraft.
>
> The Electrical, Instrument and Radio Workshop had a comprehensive array of electrical, electronic and radio equipment, as well as an assortment of aircraft instruments. The shop was also equipped with most of the test equipment that the boys would be using when they went to the hangars.
>
> Although the tool crib was well-equipped with both hand tools and power tools of various kinds, it was a condition of employment that each boy have his own tool box

and a set of basic tools. If they had a letter from Qantas, they could get a box and good quality tools at a discount from Pauls Merchants.

The boys spent the whole of their first year in classes and workshops at Rosebery and did not begin attending courses at Sydney Technical College until the beginning of their second year. The idea was to give them a full year's practical training at the School, so that when they went into one of the Company's workshops, they were capable of doing a job — the leading hand or foreman did not have to spend time showing them what to do or how to do it. This also helped to ensure that they did not pick up any bad habits from the tradesmen in the workshops. They were taught the way Qantas wanted the work to be done.'

Although Qantas trained the bulk of its apprentices to meet its own demands at the Jet Base and at Line Stations across the Qantas network, the company became involved in the late 1950s in training apprentices from countries in South-East Asia under an Australian Government initiative known as the Colombo Plan. Under this plan, apprentices from Singapore, Malaysia and the Philippines were brought to Australia for varying periods of up to six years to train as aircraft engineers at the Jet Base and the Rosebery Technical School, before returning to their own countries. And in 1971, Qantas scored an Australian aviation 'first' when it chose six young girls from forty female applicants to begin a four-year Instrument and Electrical apprenticeship at Rosebery. The girls, Kerry Mayfield (who was just fifteen years old), Annette Dries, Jeni Rindfleish, Chris Davies, Heather McPhee and Yvonne Stack (all of whom were aged sixteen), followed exactly the same training schedules as their male classmates, and they too had to pass tests in the operation of lathes, jigs and borers.

The coming of the jet era brought with it the need for Qantas Engineering to reshape its management techniques to meet the growing complexity of decision making. One of the important steps it took was to expand and strengthen its Production Planning Division by the introduction of an Industrial Engineering group under Chief Planning Engineer Arch Graham, who recalls ...

'The introduction of the 707 fleet had a dramatic impact on the workload and the demands of the relatively small Production Planning staff that existed at Mascot at that time. This was brought about by the technological explosion, which came upon us fairly rapidly. At the same time, with the complexity of the equipment, came a blowout in capital investment. This was unbelievably high and it required much more precise, scientifically managed control.

In the past, we had concerned ourselves mainly with the servicing of aircraft, determining the most economic allocation of aircraft to the various services, and the major component and overhaul workloads which were required to be carried out at each turnaround of the aircraft.

As the international airline competition intensified, we had to extend our influence to the engineering back-up shops. We also needed to have closer communication with the Corporate Planning staff, so that we were kept advised of changes in demands for the utilisation of the aircraft.

As a result, it became necessary for us to recruit a team of highly trained and qualified people with practical ability as quickly as possible.'

Arch Graham enlisted Keith Ower, Cyril Aynsley, Cliff Roberts, Ted Wickham, Aub McSweeney and Terry Quayle (all of whom had had long experience in maintenance and overhaul with Qantas) to assist him in the expansion of his Division. They were soon joined by graduate Industrial Engineers John Brady, Barry Morrow and Ted Reader, who all came to Qantas with wide experience in the field of Industrial Planning.

The Production Planning Division was divided into three main groups: Industrial Engineering, headed by Superintendent John Brady, whose Central Planning staff of twelve covered programming, budgeting, production, standards and methods; Production Planning–Mascot Hangars under Controller Keith Ower; and Production Planning–Repair and Overhaul Shop, with Controller Cyril Aynsley in charge. The Division's sixty-seven members formed 'the central nervous system of co-ordination for all Engineering, Maintenance and Overhaul facilities at Sydney Airport'. Their overall control affected approximately £50 million ($100 million) worth of assets and spares, and the work patterns of 1500 ground engineers and support staff.

The rapid explosion of technology brought with it a new language in aviation engineering circles. The term 'avionics' was coined to describe the increasingly important technology covering aircraft electronics, instruments and inertial navigation equipment. So dependent had aviation become on the accuracy of sophisticated navigation units, that Qantas, in a joint venture with the RAAF, created Australia's first 'Clean Room' at the Jet Base. The Avionics Workshop, to give it its correct name, was built at a cost of $500,000, and provided the clinical temperature-controlled and dust-free atmosphere in which delicate instrument work could be carried out accurately. It was the first laminar flow installation in Australia and worked on the principle that a steady flow of air passed through the 82 foot by 34 foot area at the rate of 100 feet per minute.

The air came into the room through a complete wall of filters from an air conditioning and temperature control plant, passing steadily across the white Laminex-topped benches where the avionics technicians worked, and up through an exhaust at the end of the room, where it became available for cleaning and re-use. The continuous flow of air swept any dust particles away, and the area immediately in front of the wall of filters achieved 'a state of cleanliness comparable only with outer space'. Keeping the air clean was not only a matter of machinery. It was also a frame of mind and a method of approach to the job. Workers in the 'Clean Room' dressed in totally enveloping dust-free nylon gowns and hoods and special footgear to maintain the sterile environment. One of the first activities undertaken in the 'Clean Room' was the servicing of the inertial navigation equipment from the RAAF's Orion ocean-patrol aircraft, and the RAAF installed test equipment valued at approximately $500,000 to enable this to be carried out.

By the end of 1967, Qantas had twelve 707-338Cs operating on its international routes, and its fleet of Boeing V-Jets now totalled twenty-two. The first three of the Qantas 707-138s (VH-EBA, VH-EBB and VH-EBC) had been sold during 1967. All the Qantas 707s had now been equipped with Doppler Navigators, which were capable of directing them automatically to their destination within the limits of the accuracy of their compasses. The Doppler system operated completely independently of ground-based aids. It used radar to measure the aircraft's ground speed and drift and fed that information into the automatic pilot, which then made the appropriate corrections to the aircraft's course. The information was also displayed to the pilot on his instruments, appearing as distance in miles to the right or left of track, and miles still to go to destination.

In the mid-1960s, Qantas had planned for its fleet of 707s to remain in service until 1970, when it was anticipated that supersonic aircraft would provide the next big advance in civil aviation. As early as 1964, the company had paid deposits to secure production-line positions for four Anglo-French Concordes and six American Boeing 733 Supersonic Transports. But as reports began to indicate that delivery of the supersonics would be several years later than expected, Qantas was faced with the realisation that their planned fleet of 707s would not provide sufficient capacity in the early 1970s, and any additional 707 purchases would become obsolete in a very short time.

Thus began an intensive study to determine whether bigger aircraft, carrying more passengers, would see Qantas through this bridging period. One of the aircraft examined was the stretched version of the Douglas DC8, which could carry up to 250 passengers. But the problems associated with operating a two aircraft-type fleet, such as the burden of extra spares and training, led Qantas to believe that the purchase of additional Boeing 707s would be a better proposition.

Then, in early 1966 Boeing announced its decision (backed by a big order from Pan American Airways) to commit its giant 747 Jumbo Jet to production, and a team of Boeing economists and engineers came to Australia to present the new aircraft to Qantas. The 747 was a civil adaption of the Boeing proposal for a C5A prototype that Boeing had designed in an unsuccessful bid to provide the United States Government with a huge military troop and cargo carrier. For Qantas, as for the other major international airlines, the Boeing 747 initiative was to change radically the pattern of world airline operations.

With a definite proposal before them, Qantas Management began a far-reaching assessment of the company's expansion up to 1980. Allowing for the possibility that supersonic aircraft could effect capacity considerably by the mid-1970s, it was concluded that Qantas would be able to introduce a high-capacity aircraft and still maintain a suitable number of flights on the key Kangaroo, Southern Cross and Tasman routes. On 2 November 1967 Qantas placed firm orders for four Boeing 747s for delivery in 1971.

The final nine 707-338s (VH-EAB *City of Canberra*, VH-EAC *City of Sydney*, VH-EAD *City of Melbourne*, VH-EAE *City of Brisbane*, VH-EAF *City of Adelaide*, VH-EAG *City of Hobart*, VH-EAH *City of Perth*, VH-EAI *City of Darwin* and VH-EAJ *City of Geelong*) were all delivered to Qantas and placed into service during 1968, replacing five of the company's 707-138Bs, which had been sold to various overseas airlines during the year.

In December 1968, Ken Gould (who had been Sydney Technical College's 'Best Apprentice Ground Engineer' in 1957) began what was to be a four-year posting to Seattle as Senior Qantas Technical Representative at Boeing's new Everett plant. His main responsibility was to ensure that the Qantas specifications were adhered to in the construction and subsequent delivery of the company's first four Boeing 747s. Ken Gould recalls ...

> 'The interesting thing about the 747 was the fact that all of Boeing's customers for the new aircraft, including Qantas, had bought 'paper aeroplanes' — there was no prototype at the time initial orders for the 747 had been placed.
>
> By February 1969, however, Boeing had built and test-flown its first 747 and Mick Ryan, who was with me in Seattle at that time, and I spent most of 1969 working through the certification programme for the aircraft. Boeing would hold a meeting every Thursday to advise the Technical Representatives of the customer airlines who were stationed in Seattle, of developments (both positive and negative) in the 747 programme, and we would communicate this information in weekly progress reports to our Management back in Sydney.
>
> The first Qantas 747 was allocated production line number 147. Qantas had produced about three hundred changes to the original Boeing specification for the 747, to ensure its compatibility with the manner in which we proposed to operate the aircraft. For example, we were the first airline to get what was termed a 'mini extended upper deck' — our 747 was twenty inches longer in the upper deck, which allowed the window configuration in that area of the aircraft to be increased from the standard three to ten on each side. Ours was the first 747 to have ten windows.'

Joseph Sutter, a Senior Boeing Executive who has now retired but remains as a consultant to the organisation, was recently moved to make the following observations about Qantas:

> 'I was Chief Engineer on the 747, and I stayed with the programme full-time for about ten years.
>
> I think Qantas took a very bold step for a small airline to decide to compete with the Pan American's, the TWA's and the Lufthansa's. It took a lot of courage for Qantas to launch the 747 in that area of the world.
>
> They not only knew what they wanted though, they also knew what they needed, and they were not shy about asking for it. Ken Gould was their representative here, and he came to my office many times to tell me what he didn't like and what he wanted.
>
> Because we came to know that Qantas did know what they needed and also knew how to tackle any problems which arose, we paid a lot of attention to Qantas.'

Meanwhile, back in Sydney, the Australian Government had begun construction of a new International Terminal at Mascot, and the extension of the North–South runway into Botany Bay. However, the terminal facilities had been designed in 1963 to handle the British and American supersonic aircraft and subsonic narrow body types such as the Boeing 727, which had begun operation on Australia's domestic routes. Now, with the introduction of the 747s scheduled for 1971, planning for handling facilities at the new terminal had to undergo substantial modification, and a further extension of the runway into Botany Bay became a priority.

When the first Boeing 747 rolled out of the Boeing Everett factory complex on 30 September 1968, onlookers described the size of the aircraft as 'awesome'. And no less awesome were the challenges facing the Qantas engineers at Mascot, as they planned for the introduction of the Jumbo Jets.

Relocation of the International Terminal to the western side of the airport meant that aircraft had to be towed to and from the Jet Base across the North–South runway. To fulfil this task, the Fox Manufacturing Company of Sydney submitted a design to Qantas for a tug capable of towing 400 tons at 10 miles per hour. Powered by twin V8 diesel engines, the tug served a dual purpose: the economic towing of both 747s and 707s. A unique design feature allowed the driver to manoeuvre backward and forward with full visibility; his seat, steering wheel and instrument console could be swivelled 180 degrees to face the direction he was travelling. Qantas ordered eight Fox tugs for use at the Jet Base in Sydney and at its Line Stations.

Part of the Engineering Department's gearing-up for the Boeing 747B involved the establishment of a night shift in March 1970. Confined initially to the Base Servicing Group, the introduction of a 24 hour day, seven-day-a-week shift pattern resulted in aircraft being available to return to service much sooner than previously.

Construction of a new 747B hangar complex with a two-storey maintenance annex was started in 1970, as special equipment for handling and servicing the 747s began to appear at the Jet Base. Five 62 foot high aerial work platforms, commonly known as 'cherry-jet pickers', were specially designed to carry out servicing and maintenance of the 747Bs. Twelve Jumbo container-pallet loaders, which operated on a scissor-type platform in the loading and unloading of cargo on the 747Bs, were placed on order with FMC (Australia) in Victoria, and Hastings Deering in Sydney again provided the passenger loading steps for the new aircraft. The 747B steps were more revolutionary than those used on the 707s, being completely enclosed from the wind and rain, and with a special access provided for wheelchairs.

An automatic stacker-retriever was installed at the Central Store as part of the expansion program for the Procurement and Supply Department's facilities at the Jet Base. The new concept in warehousing comprised 66 foot high rows of pallet racing, which held spares and equipment in pigeon-holes. Cranes operated by a punch-card system from a single console ran between the rows selecting stock automatically.

As the delivery date for the first Qantas 747B drew near, air crews began intensive training courses in the 747B Flight Simulator, which had been installed at the Jet Base. And engineering staff went to the Pratt and Whitney Training School at Hartford, Connecticut, to study the giant JT9D-3A engines that were to power the first four Qantas Jumbos. In May 1971 the first of Qantas' spare JT9D engines arrived at the Jet Base and was used for training engineering and maintenance staff in Sydney.

Amidst all the preparation for the arrival of the Boeing 747Bs, Qantas celebrated its fiftieth anniversary. To commemorate the occasion a special display was mounted in the original Qantas hangar at Longreach, outlining the company's fifty-year history, from the Avro 504K to the Boeing 707-338Cs. It was a time of reflection for Engineering Facilities Manager Jack Avery. He had been the first apprentice to be employed by Qantas and had begun his long and distinguished career with the company in the hangar at Longreach in 1927. At the time of his return to Longreach for the fiftieth anniversary celebrations, he was the company's longest serving employee, having completed a total of forty-three years. It was also a time of great pride for Jack Avery as the Qantas Chairman, Sir Roland Wilson, presented him to Queen Elizabeth and Prince Philip during their visit to the exhibition.

In February 1971, Qantas General Manager, Captain R.J. (Bert) Ritchie announced a top level reorganisation in which Ron Yates became the company's first Director of Engineering and Maintenance. (Ron Yates' career with Qantas continued its upward spiral with his appointment in 1977 to the position of Deputy General Manager and Chief Operating Officer. His services to the aviation industry were recognised in 1983 by his being awarded the honour of becoming a Member of the Order of Australia. Then in January 1985, in what was to be the culmination of his career with the company, he became the first engineer to be appointed to the position of Chief Executive Officer of Qantas.)

At 7:40 am on Monday 16 August 1971 the first Qantas 747-238B (VH-EBA, named *City of Canberra*) touched down at Sydney's Kingsford-Smith Airport, escorted by the company's two HS-125 pilot trainers. By the end of 1971, Qantas had taken delivery of four Boeing 747Bs and had placed an order for a fifth.

A 1993 PERSPECTIVE

A 747 decision proved to be a good one for Qantas. During the next twenty years the company would finally dispose of the last of its propeller-driven aircraft to concentrate exclusively on its rapidly expanding 747 fleet. (Two Qantas DC3s, VH-EDC and VH-EDD, were sold in late 1971 shortly after the first 747Bs entered service. Then in August 1977 the two remaining Qantas DC4s, VH-EDA and VH-EDB, were purchased by Air Express Limited.) Qantas introduced a variety of 747 models and engine types during this period, and in the early 1980s it became the only airline in the world to operate an 'all-747' fleet.

The Qantas Jet Base at Mascot came to be recognised as a world-class engineering facility, and in 1985 it achieved what is possibly the greatest recognition of its engineering skill when Boeing endorsed it as the only facility in the world, outside its own plants in the United States, authorised to carry out airframe modification work on 747 aircraft.

For more than seventy years, Qantas engineers have led the world in achieving standards of technological excellence — standards based on the 'tradition of integrity' pioneered by the company's first engineer, Wilfred Arthur Baird, in the old corrugated-iron hangar at Longreach during the 1920s. In early 1993, the Qantas Jet Base at Mascot was renamed the Qantas 'Arthur Baird' Jet Base — a formal recognition of his singular contribution to the development of Qantas as one of the world's most respected airlines.

INDEX